Visual
Category
Theory

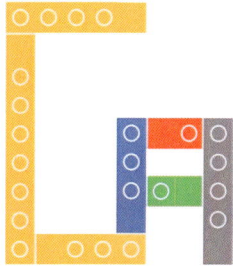

Dmitry
Vostokov

brick by brick

$\{Part_0, Part_1..., Part_7\} \rightarrow C^{Parts}$

Visual Category Theory Brick by Brick: LEGO® Reference

Published by OpenTask, Republic of Ireland

OpenTask books and magazines are available through booksellers and distributors worldwide. For further information or comments, send requests to press@opentask.com.

A CIP catalog record for this book is available from the British Library.

ISBN-13: 978-1912636389 (Paperback)

Revision 1.00 (October 2021)

Dmitry Vostokov is an internationally recognized expert, speaker, educator, scientist, and author. He is the founder of the pattern-oriented software diagnostics, forensics, and prognostics discipline and Software Diagnostics Institute (DA+TA: DumpAnalysis.org + TraceAnalysis.org). Vostokov has also authored more than 50 books on software diagnostics, anomaly detection and analysis, software and memory forensics, root cause analysis and problem solving, memory dump analysis, debugging, software trace and log analysis, reverse engineering, and malware analysis. He has more than 25 years of experience in software architecture, design, development, and maintenance in various industries, including leadership, technical, and people management roles. Dmitry also founded Syndromatix, Anolog.io, BriteTrace, DiaThings, Logtellect, OpenTask Iterative and Incremental Publishing (OpenTask.com), Software Diagnostics Technology and Services (former Memory Dump Analysis Services) PatternDiagnostics.com, and Software Prognostics. In his spare time, he presents various topics on Debugging.TV and explores Software Narratology, its further development as Narratology of Things and Diagnostics of Things (DoT), and Software Pathology. His current interest areas are theoretical software diagnostics and its mathematical and computer science foundations, application of formal logic, artificial intelligence, machine learning and data mining to diagnostics and anomaly detection, software diagnostics engineering and diagnostics-driven development, diagnostics workflow and interaction. Recent interest areas also include cloud native computing, security, automation, functional programming, and applications of category theory to software development and big data.

Visual Category Theory

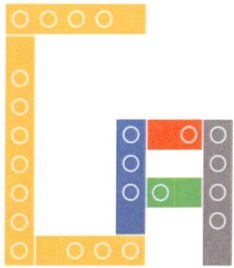

Dmitry
Vostokov

brick by brick

$$Part_0 \in C^{Parts} | \forall Part_x \exists! : Part_0 \rightarrow Part_x$$

Visual Category Theory Brick by Brick, Part 0: Using LEGO® to Teach Abstract Mathematics

Published by OpenTask, Republic of Ireland

OpenTask books and magazines are available through booksellers and distributors worldwide. For further information or comments, send requests to press@opentask.com.

A CIP catalog record for this book is available from the British Library.

ISBN-13: 978-1912636396 (Paperback)

Revision 1.01 (April 2021)

Preface

This free to download and distribute Part 0 was added to Visual Category Theory Brick by Brick series to explain mathematical notation and other foundational aspects used in brick construction annotations. Some readers expressed concern that the notation used is unfamiliar to them, and I realized that not everyone is continuously reading books on mathematical logic and set theory or studied proper foundations of mathematical analysis where they encountered logical quantifiers. Readers who already bought the Visual Category Theory bundle from other sources other than directly from me can download this part if necessary. The cover subtitle alludes to an initial object in category theory introduced in Part 2.

The choice of object notation for sets in this part matches the notation in subsequent parts.

The short Handbook of Logic and Proof Techniques for Computer Science by Steven G. Krantz is recommended for further studying or filling the gaps. In addition to topics from the book title, it covers many others, including elementary and axiomatic set theories, recursive functions, lambda calculus, groups, Boolean algebra, and complexity theory. It also has a short chapter on category theory.

Information about further parts, including sample pages and index, can be found on the Software Diagnostics Institute page:

https://www.dumpanalysis.org/visual-category-theory

Contents of Parts 0 - 7

Objects (elements) of a set *S* (collection of objects) are chosen from some Universe *U*: $S = \{X, Y, Z, P\}$

U

S

X

Y

Z

P

Set-builder notation (|): $A = \{X | X \text{ is red}\}$ and $B = \{X | X \text{ is blue}\}$

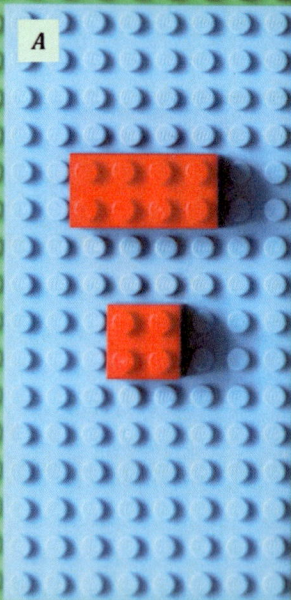

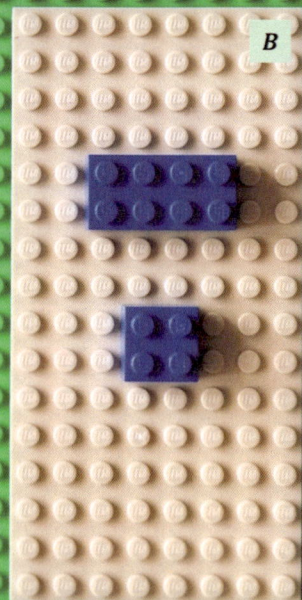

A

B

Another example: $C = \{X | X \text{ is square}\}$ and $D = \{X | X \text{ is green}\}$

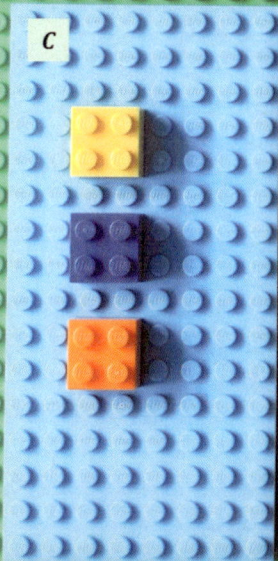

C

D

Set membership ($\in$): $X \in A$, $Y \in B$, $Z \notin A$, $Z \notin B$

(Proper) subset inclusion ($\subset, \supset$): $A \subset B$ and $B \supset A$ but $B \nsubseteq A$
All ($\forall$) elements of A are elements of B: $\forall X \in A, X \in B$
There are some ($\exists$) elements of B are not elements of A: $\exists Y \in B, Y \notin A$

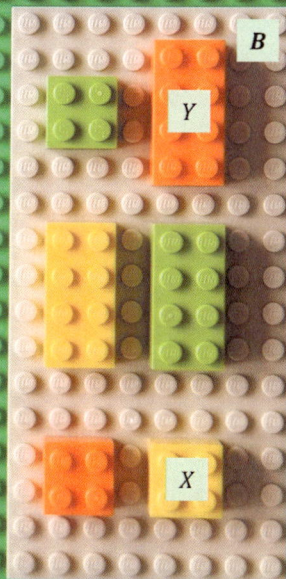

Not a proper subset inclusion ($\subseteq, \supseteq$): $A \subseteq B$ and $A \supseteq B$ because $A = B$

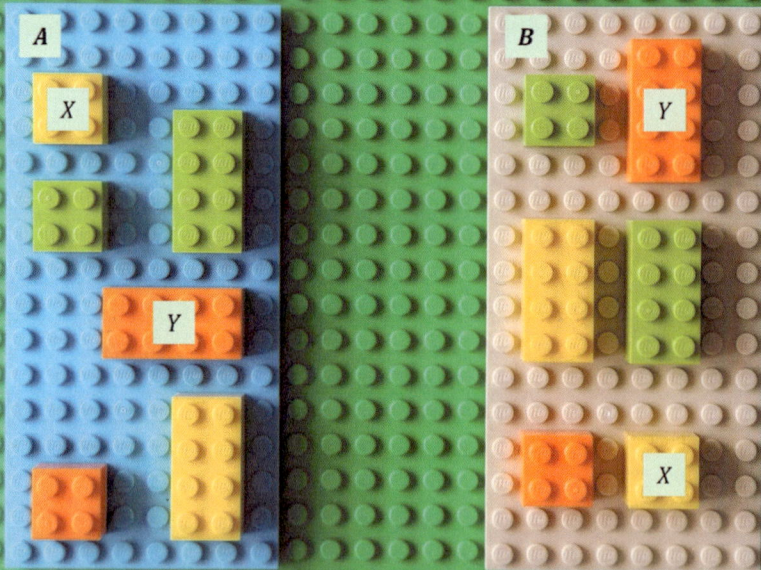

Sets can be members of sets: $D \in B$
Set as a member is not a subset here: $D \nsubseteq B$ because $\exists Y \in D, Y \notin B$
$D = \{X, Y\}$
$B = \{Z, X, D\} = \{Z, X, \{X, Y\}\}$

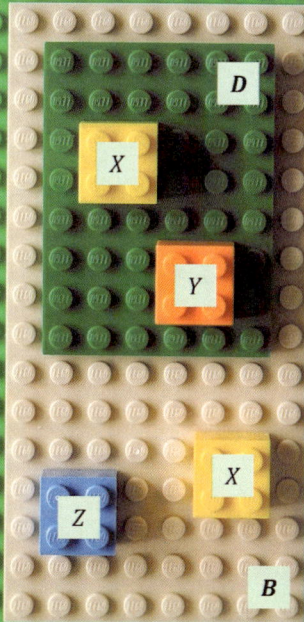

Another example of elements and sets as members:

$X \in A$ and $X \in B$ and $X \notin C$ and $X \in D$
$Y \notin A$ and $Y \in B$ and $Y \notin C$ and $Y \notin D$
$Z \notin A$ and $Z \notin B$ and $Z \in C$ and $Z \notin D$
$W \notin A$ and $W \notin B$ and $W \in C$ and $W \notin D$
$B \in A$ and $C \in A$ and $D \in A$
$B = \{X, Y\}$ and $C = \{Z, W\}$ and $D = \{X\}$
$A = \{X, B, C, D\} = \{X, \{X, Y\}, \{Z, W\}, \{X\}\}$

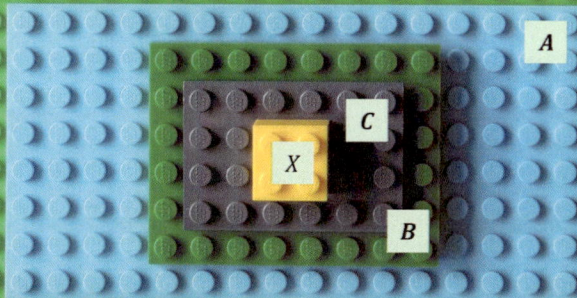

Sets can be nested:
$X \notin A$ and $X \notin B$ and $X \in C$
$B \in A$ and $C \in B$ but $C \notin A$
$C = \{X\}$
$B = \{C\} = \{\{X\}\}$
$A = \{X, B\} = \{X, \{C\}\} = \{X, \{\{X\}\}$

The difference between membership and subset:
$B = \{X\}$
$A = \{B\} = \{\{X\}\}$
$B \in A$ but $B \not\subseteq A$ because $X \notin A$

The powerset $P(S)$ of a set S is the set of all subsets
of S including S itself and the empty set $\emptyset$ (the
proper subset of every set)
$S = \{X, Y\}$
$P(S) = \{\emptyset, \{X\}, \{Y\}, \{X, Y\}\}$

$P(S)$

$\emptyset \subset S$

$\{Y\} \subset S$

Y

$S \subseteq S$

X   Y

$\{X\} \subset S$

X

A relation between two sets is a collection of ordered pairs from each set:

$A \sim_R B = \{(X,P),(X,T),(Y,R),(Y,Q)\}$

Each source member may map to several target members, for example, X to P and T

A

(X,P)

(X,T)

X

Y

(Y,R)

(Y,Q)

B

P

Q

R

T

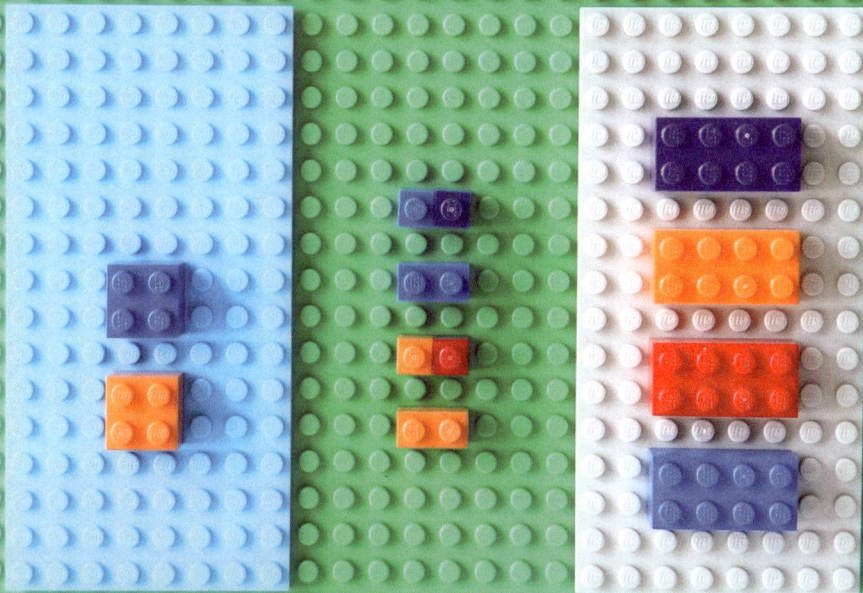
A compact way to show individual relation mappings

A functional relation (partial function) between two sets is a collection of ordered pairs from each set where for each source member, there is at most one target member:

$$A \sim_f B = \{(X,T),(Y,Q)\}$$

The most frequent functional relation notation:

$$f : A \rightarrow B$$

Another example of a functional relation $f : A \to B$
$A = \{X, Y\}$ is the domain of the function f
$B = \{P, Q, R, T\}$ is the codomain of the function f
$\{P, T\}$ is the range of the function f

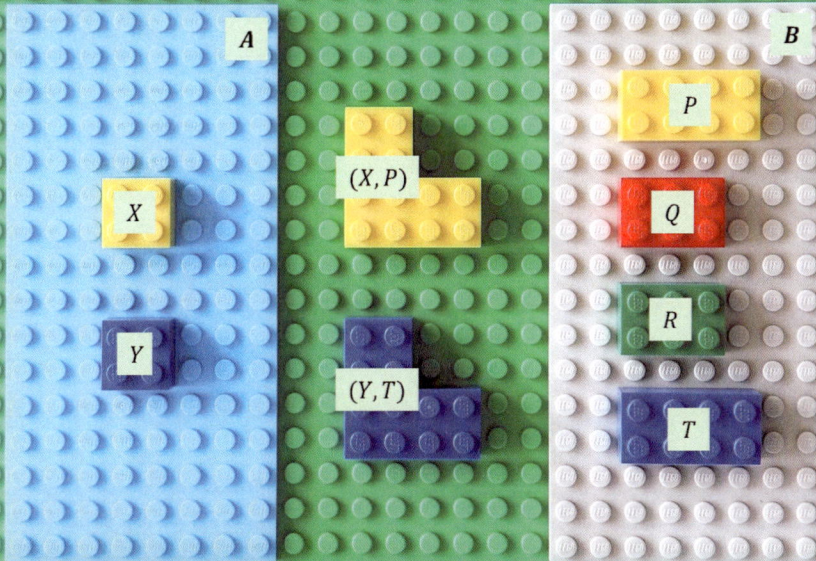

A

B

(X, P)

X

Y

(Y, T)

P

Q

R

T

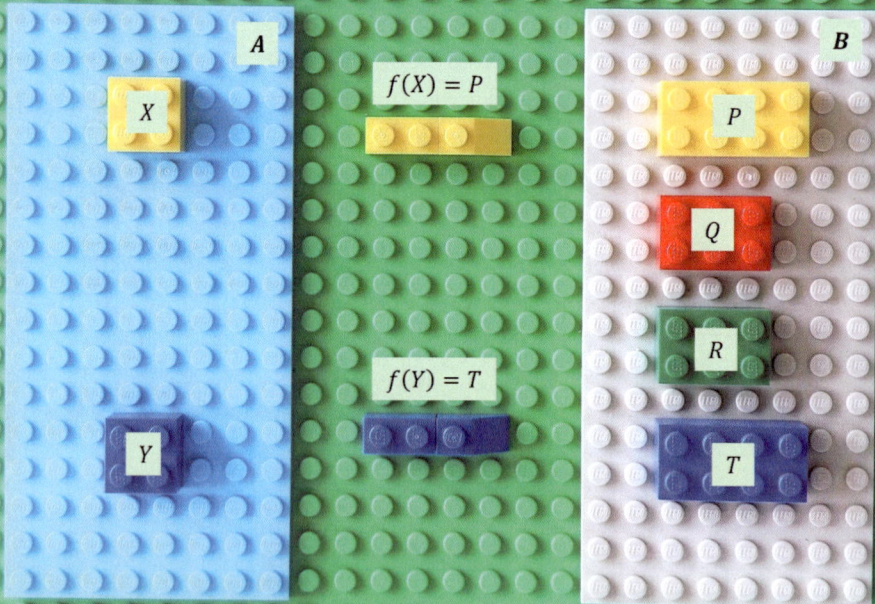

A compact way to show individual functional mappings for the function $f: A \to B$ using brick arrows

A

X

$f(X) = P$

Y

$f(Y) = T$

B

P

Q

R

T

A function $f: A \to B$ is injective (one-to-one, injection) if distinct elements from the domain A are mapped to distinct elements from the codomain B:
if $X \neq Z$ then $f(X) \neq f(Z)$

A

X

Z

Y

$f(X) = P$

$f(Z) = Q$

$f(Y) = T$

B

P

Q

R

T

This function $f: A \to B$ is **not** injective because there are elements from the domain A that are mapped to the same element from the codomain B: $X \neq Z$ but $f(X) = f(Z)$

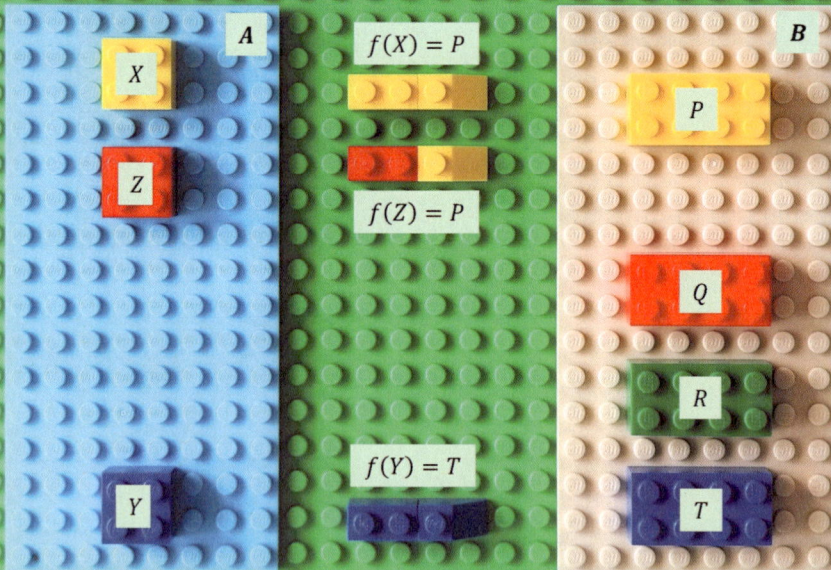

A

X

Z

Y

$f(X) = P$

$f(Z) = P$

$f(Y) = T$

B

P

Q

R

T

A function $f: A \to B$ is surjective (onto, surjection) if every ($\forall$) element from the codomain B is mapped to by at least one ($\exists$) element from the domain A:
$\forall elem_B \in B, \exists elem_A \in A, f(elem_A) = elem_B$
The function depicted here is surjective but **not** injective

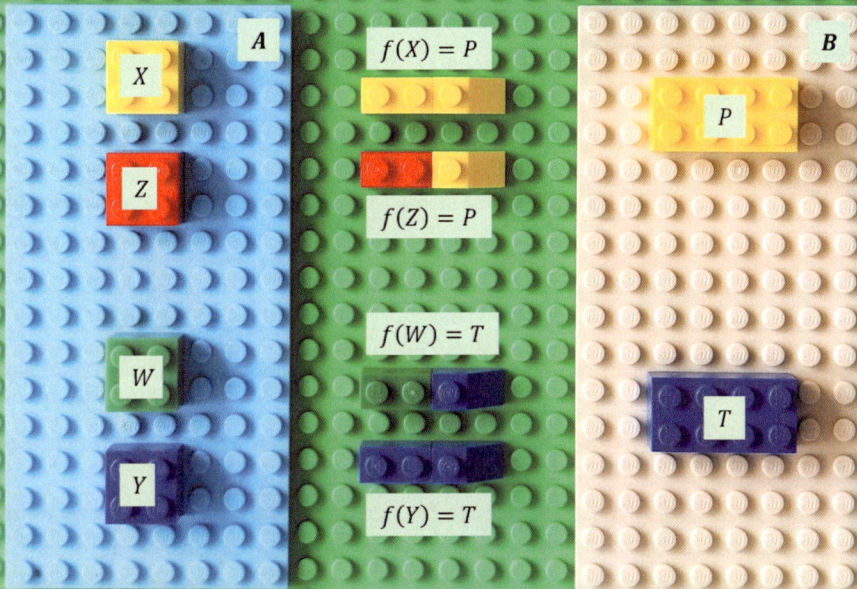

A

X

Z

W

Y

$f(X) = P$

$f(Z) = P$

$f(W) = T$

$f(Y) = T$

B

P

T

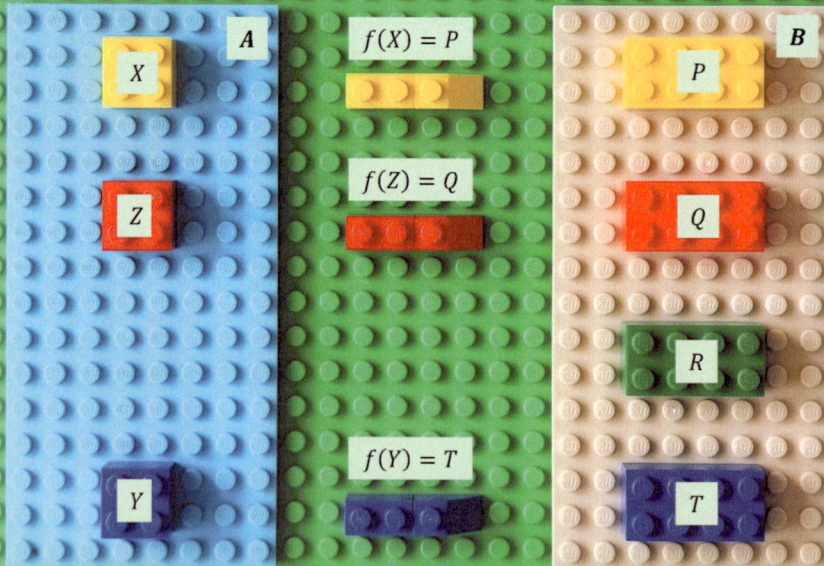

The function depicted here is injective but **not** surjective

A

$f(X) = P$

$f(Z) = Q$

$f(Y) = T$

B

X

Z

Y

P

Q

R

T

A function $f: A \to B$ is bijective (one-to-one and onto, invertible, bijection) if every ($\forall$) element from the codomain B is mapped to by exactly one ($\exists!$) element from the domain A:

$$\forall elem_B \in B, \exists! \, elem_A \in A, f(elem_A) = elem_B$$

A product of two sets A and B is a set $A \times B$ of all ordered pairs $(elem_A, elem_B)$ where $elem_A \in A$ and $elem_B \in B$

$A = \{X, Y\}$

$B = \{P, Q\}$

$A \times B = \{(X, P), (X, Q), (Y, P), (Y, Q)\}$

The order of the sets in a product is important:
$B \times A$ is different from $A \times B$
$A = \{X, Y\}$
$B = \{P, Q\}$
$B \times A = \{(P,X),(Q,X),(P,Y),(Q,Y)\}$

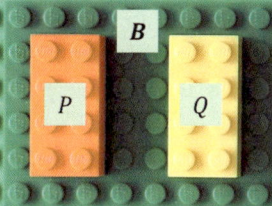

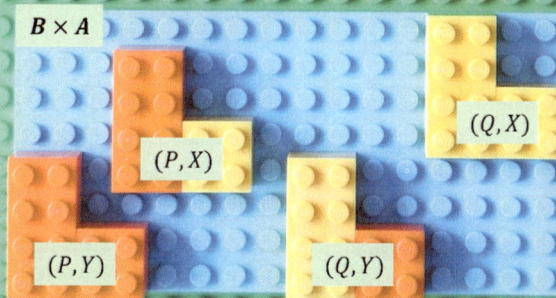

B

P   Q

A

X

Y

$B \times A$

(P,X)   (Q,X)

(P,Y)   (Q,Y)

The example of a product $A \times A$:
$A = \{X, Y\}$
$A \times A = \{(X, X), (Y, X), (X, Y), (Y, Y)\}$

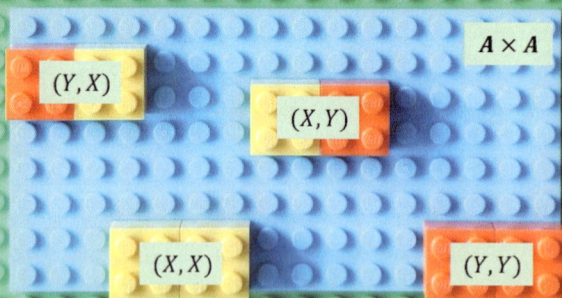

A union of sets A and B is a set $A \cup B$
with elements from A or B:
$A = \{X, Y\}$
$B = \{Y, Z\}$
$A \cup B = \{X, Y, Z\}$

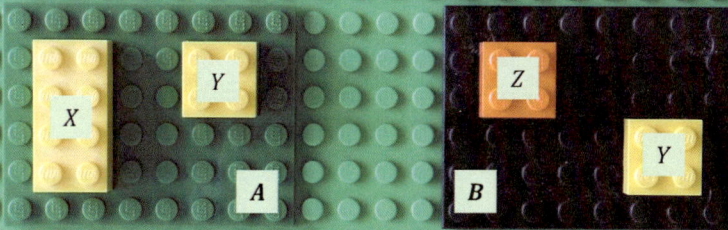

An intersection of sets A and B is a set $A \cap B$
with elements that belong to both A and B:
$A = \{X, Y\}$
$B = \{Y, Z\}$
$A \cap B = \{Y\}$

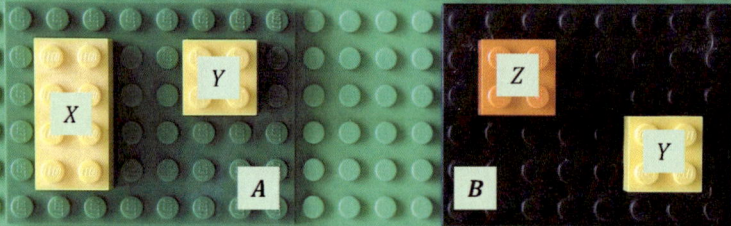

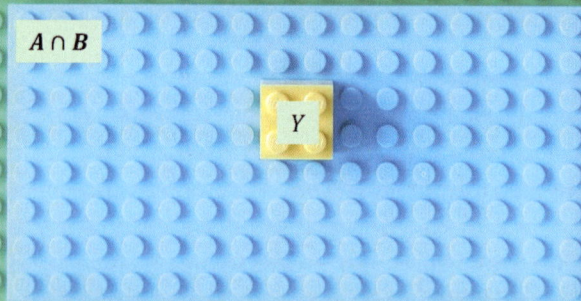

A set difference of two sets A and B is a set $A\backslash B$ or $A - B$ with elements from A that do not belong to B:

$A = \{X, Y\}$

$B = \{Y, Z\}$

$A\backslash B = \{X\}$

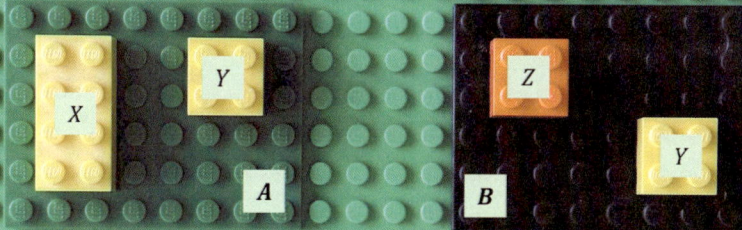

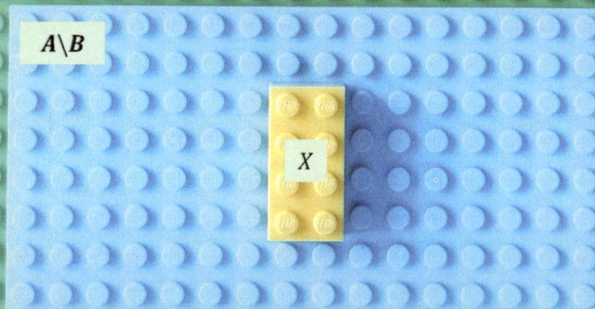

A symmetric set difference of two sets A and B is a set $A \triangle B$ with elements that belong to either A or B but not to both:
$A = \{X, Y\}$
$B = \{Y, Z\}$
$A \triangle B = \{X, Z\}$

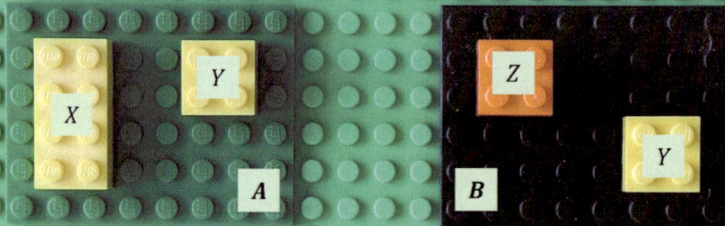

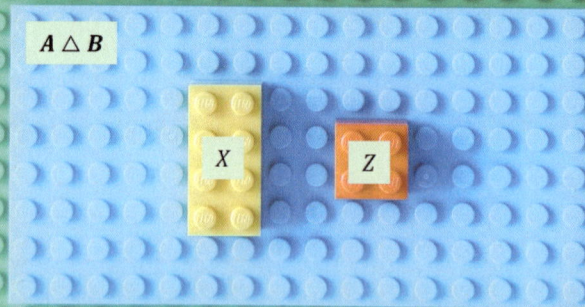

Consider two functions f and g with domain A and codomain B but different ranges:
$f(y) = b$
$f(Y) = B$
$g(y) = r$
$g(Y) = R$

These functions f and g belong to the set $\{f,g\}$ of functions $A \to B$ which is a subset of the set of all functions from A to B:

$f,g \in B^A$

$\{f,g\} \subset B^A$

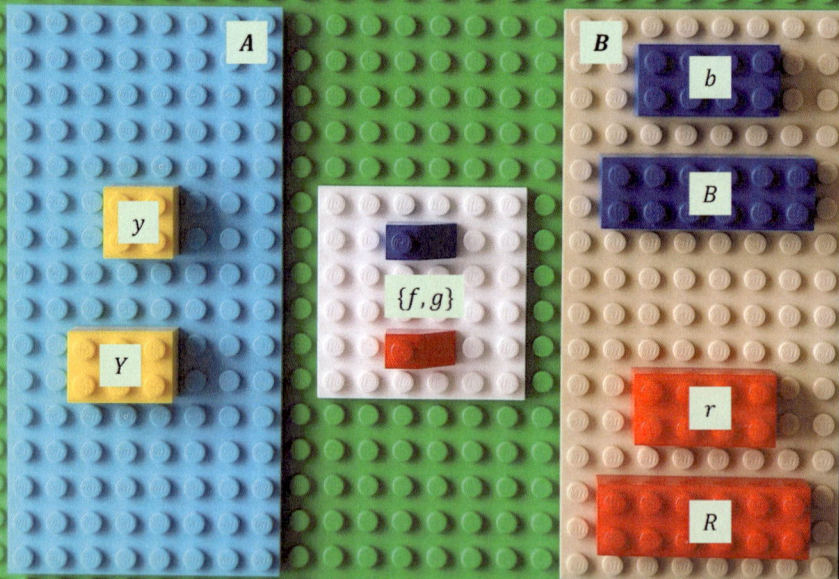

A

y

Y

$\{f,g\}$

B

b

B

r

R

A function composition ∘ of functions g and f that agree on domain/codomain (the domain of g is the codomain of f) produces a new function $h = g \circ f$ with the domain of f and the codomain of g:

$f(X) = Y$
$g(Y) = Z$
$h(X) = g(f(X)) = Z$

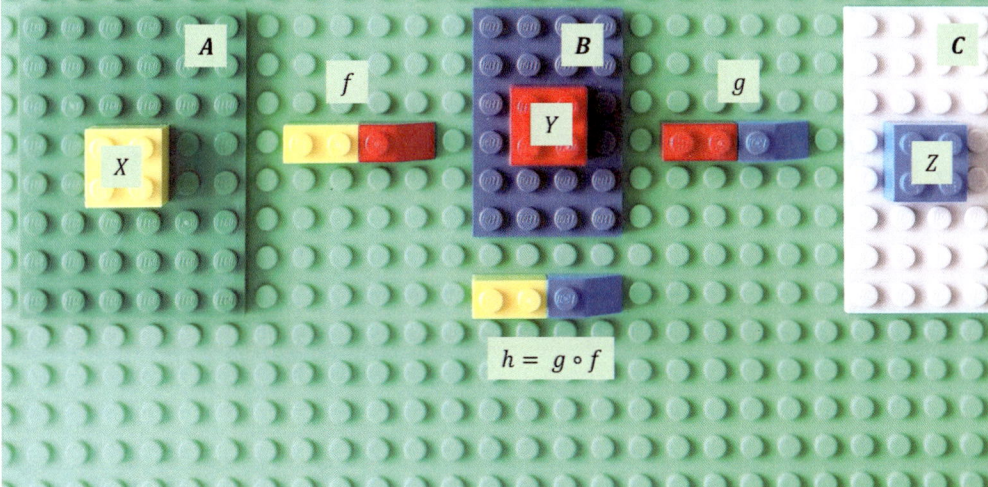

For invertible functions, we have the following inverse function composition $h^{-1} = f^{-1} \circ g^{-1}$:
$f(X) = Y$ and $f^{-1}(Y) = X$
$g(Y) = Z$ and $g^{-1}(Z) = Y$
$h(X) = g(f(X)) = Z$ and $h^{-1}(Z) = f^{-1}(g^{-1}(Z)) = X$

$h^{-1} = f^{-1} \circ g^{-1}$

A

f^{-1}

B

g^{-1}

C

X

Y

Z

f

g

$h = g \circ f$

Visual Category Theory

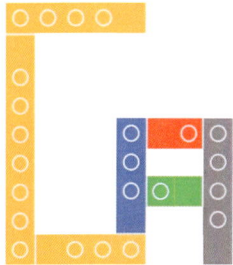

Dmitry Vostokov

brick by brick

$Part_1 \in C^{Parts}$

Visual Category Theory Brick By Brick, Part 1: Using LEGO® to Teach Abstract Mathematics

Published by OpenTask, Republic of Ireland

OpenTask books and magazines are available through booksellers and distributors worldwide. For further information or comments, send requests to press@opentask.com.

A CIP catalog record for this book is available from the British Library.

ISBN-13: 978-1912636402 (Paperback)

Revision 1.01 (April 2020)

Preface

Category theory abstractions are very challenging to apprehend correctly, require a steep learning curve for non-mathematicians, and, for people with traditional naïve set theory education, a paradigm shift in thinking. When reading various category theory textbooks, I found the lack of visual color examples, and, almost 3 years ago, in May 2017, I began working on a coffee table book. I made a few square slides shown below on this page, but then life got me carried away from writing. Recently, I started using LEGO® to teach machine learning and associated data structures and algorithms and found a way to represent directed graphs where I had to use arrows. Success with such representation struck me, and I realized that I could resume writing my previous visual category theory book but now using bricks instead of abstract circles.

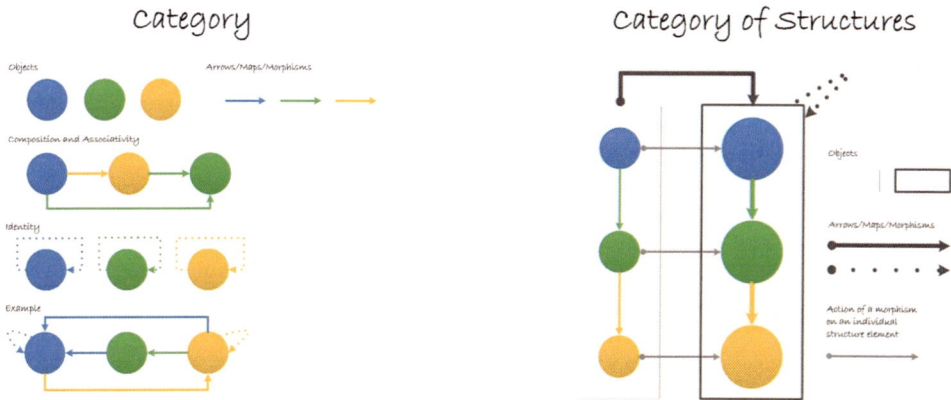

For this part, I used the following books as a reference:
- An Introduction to Category Theory by Harold Simmons
- Category Theory and Applications: A Textbook for Beginners by Marco Grandis
- Modern Classical Homotopy Theory by Jeffrey Strom

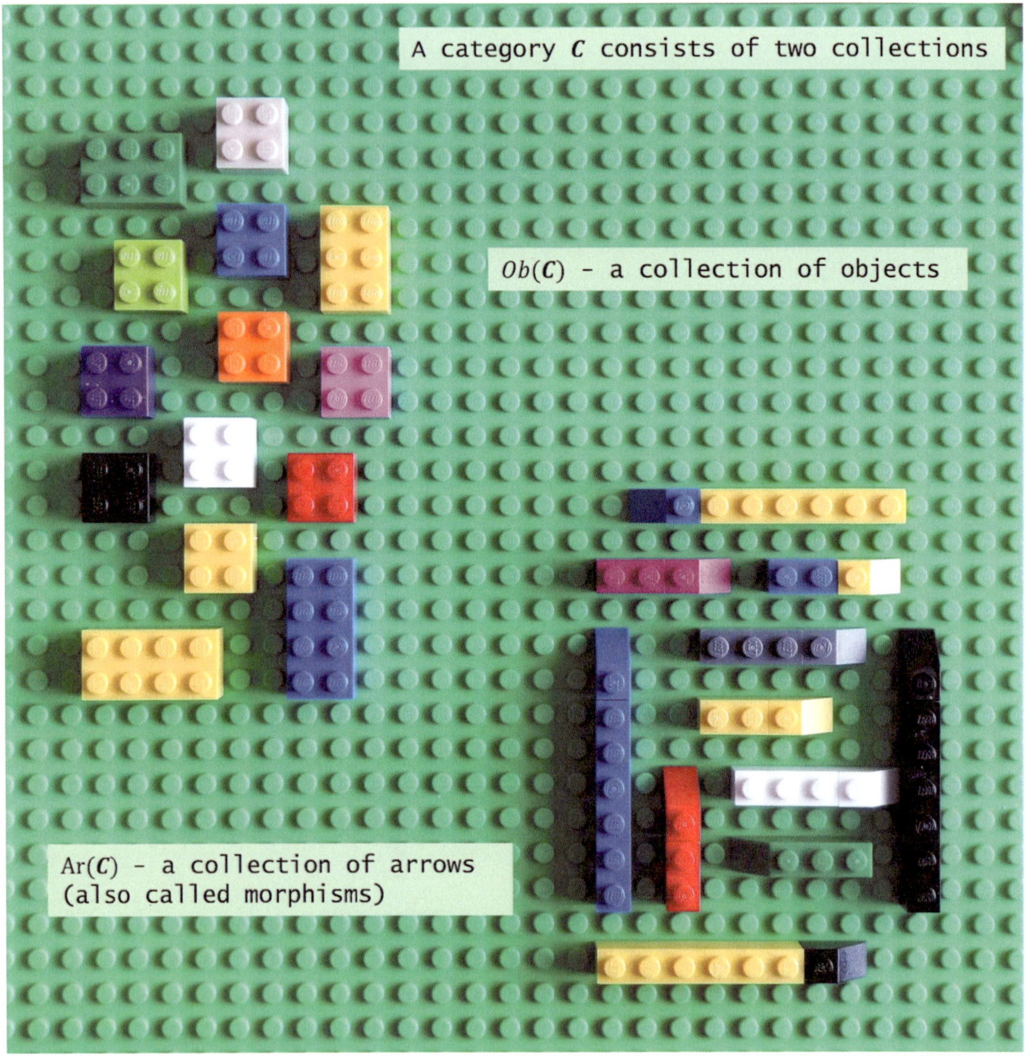

A category C consists of two collections

$Ob(C)$ - a collection of objects

$Ar(C)$ - a collection of arrows
(also called morphisms)

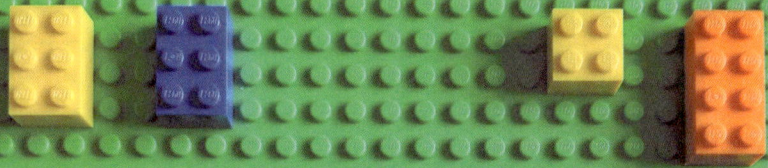

Any $X, Y \in Ob(\mathbf{C})$ have a set of arrows (maps, morphisms) between them which can be empty

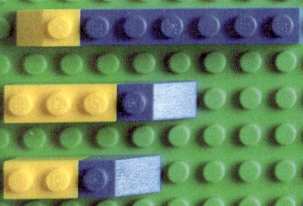

Each arrow has source (domain) and target (codomain) objects

$X, Y, Z \in Ob(\mathbf{C})$
$f, g \in Ar(\mathbf{C})$
$f \in Ar(X, Y)$
$g \in Ar(Y, Z)$

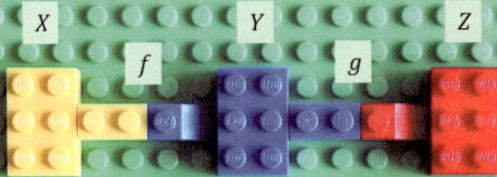

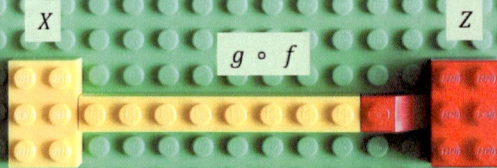

The composition of arrows is associative: $f \circ (g \circ h) = (f \circ g) \circ h$

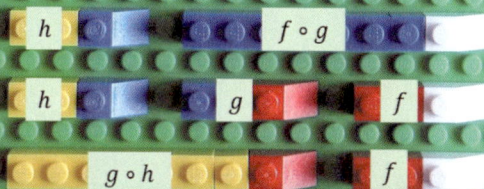

For each $X \in Ob(\mathbf{C})$ there is a special arrow $id_X \in Ar(X,X)$ that satisfies the following properties: $f \circ id_X = f$ and $id_X \circ g = g$

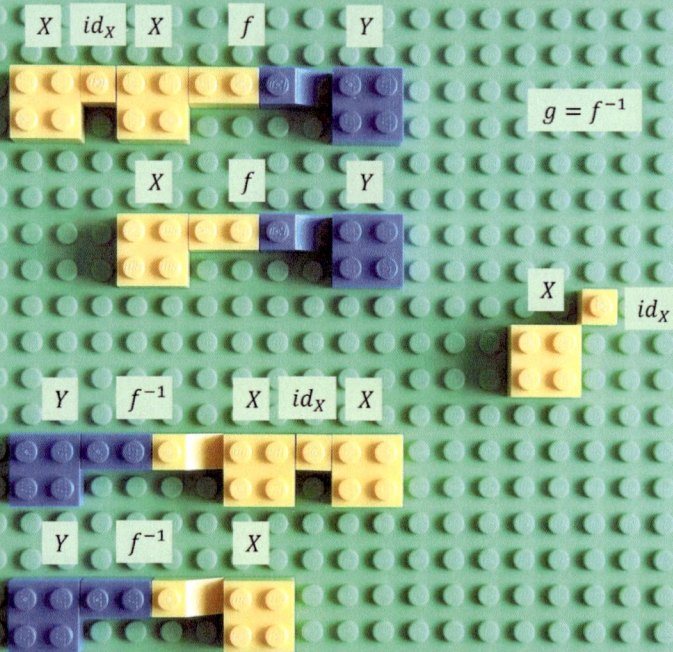

$X \quad id_X \quad X \qquad f \qquad Y$

$g = f^{-1}$

$X \qquad f \qquad Y$

X

id_X

$Y \quad f^{-1} \quad X \quad id_X \quad X$

$Y \quad f^{-1} \quad X$

$R \in Ob(\boldsymbol{C})$ is a retract of $X \in Ob(\boldsymbol{C})$
$s, r \in Ar(\boldsymbol{C})$ are section and retraction

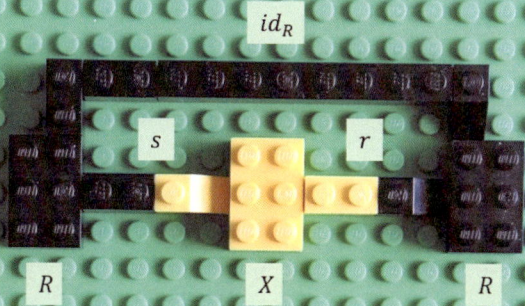

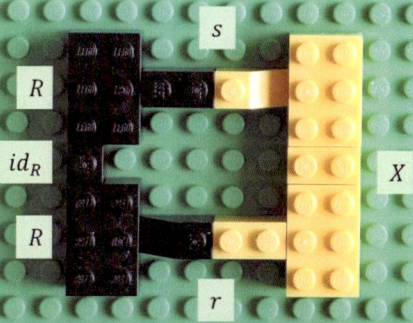

Part 1 page 9

$Ar(\pmb{C}) \ni f \colon X \to Y$ is an equivalence if there is
$Ar(\pmb{C}) \ni g \colon Y \to X$ with $g \circ f = id_X$ and $f \circ g = id_Y$

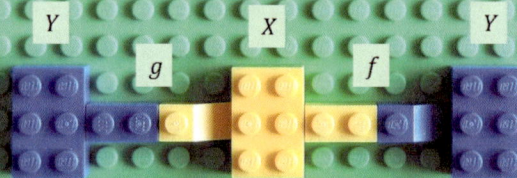

A covariant functor between categories $F: \mathbf{C} \to \mathbf{D}$
$F: Ob(\mathbf{C}) \to Ob(\mathbf{D})$ and
$F: Ar(\mathbf{C}) \to Ar(\mathbf{D})$ with $F: Ar(X,Y) \to Ar(F(X),F(Y))$

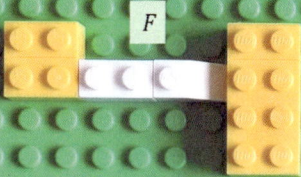

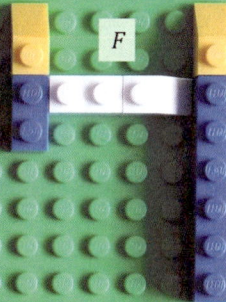

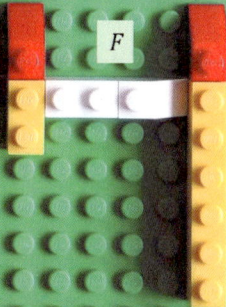

A contravariant functor between categories $G: \boldsymbol{C} \to \boldsymbol{D}$
$G: Ob(\boldsymbol{C}) \to Ob(\boldsymbol{D})$ and
$G: Ar(\boldsymbol{C}) \to Ar(\boldsymbol{D})$ with $G: Ar(X,Y) \to Ar(G(Y),G(X))$

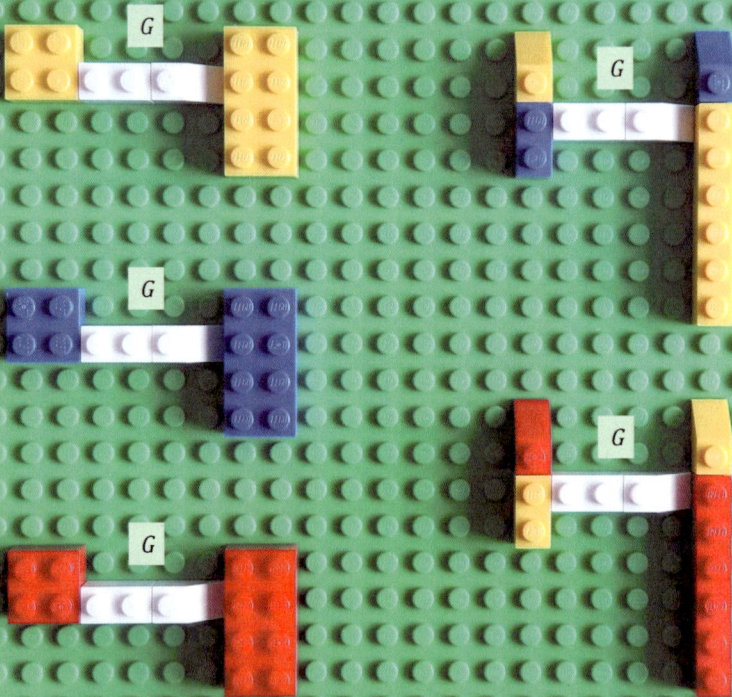

A covariant functor F and contravariant functor G

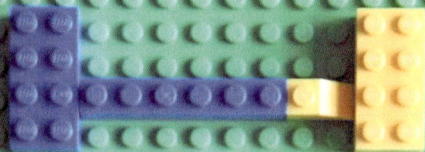

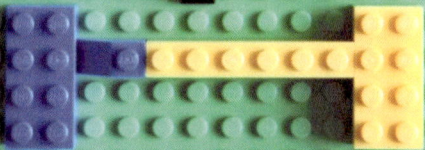

A covariant functor F and contravariant functor G
$F(g \circ f) = F(g) \circ F(f)$ and $G(g \circ f) = G(f) \circ G(g)$

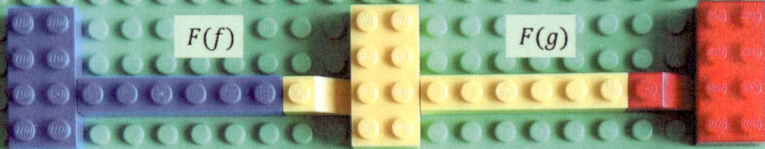

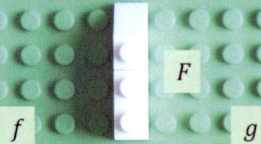

$F(f)$ $F(g)$

F

f g

G

$G(f)$ $G(g)$

A natural transformation N is a collection of arrows between functors' targets indexed by functors' source objects
$n_X \in Ar(F(X), G(X))$

A 2-category is a category with arrows having a category structure: arrows as objects (1-cells) and arrows (2-cells) between 1-cells

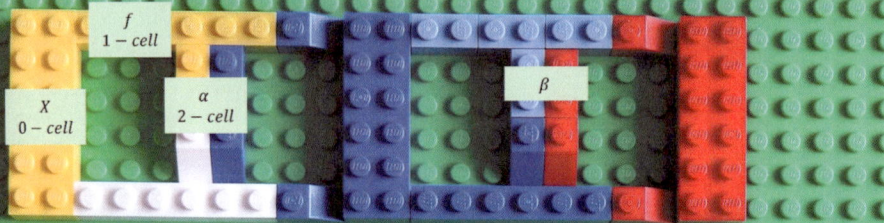

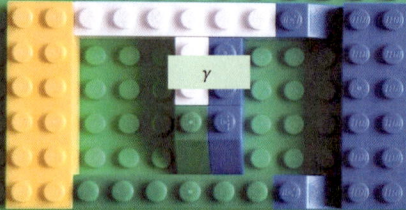

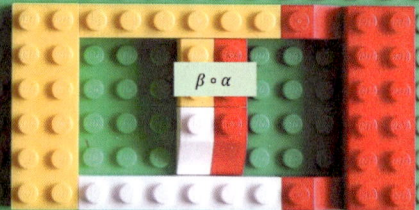

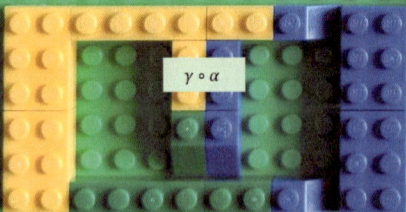

f
$1 - cell$

X
$0 - cell$

α
$2 - cell$

β

γ

$\beta \circ \alpha$

$\gamma \circ \alpha$

Horizontal composition $\circ_0$ of 2-category arrows along 0-cells

Vertical composition $\circ_1$ of 2-category arrows along 1-cells

Visual Category Theory

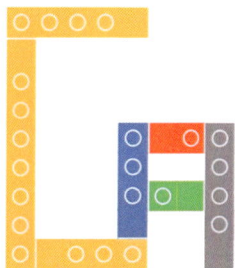

Dmitry Vostokov

rick by rick

Part$_2 \in C^{Parts}$

Visual Category Theory Brick by Brick, Part 2: Using LEGO® to Teach Abstract Mathematics

Published by OpenTask, Republic of Ireland

OpenTask books and magazines are available through booksellers and distributors worldwide. For further information or comments, send requests to press@opentask.com.

A CIP catalog record for this book is available from the British Library.

ISBN-13: 978-1912636419 (Paperback)

Revision 1.01 (June 2021)

Duality: reversing all arrows and composition order

Lifting: we lift $f = p \circ l$ to a larger target X

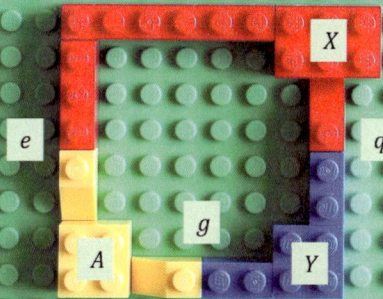

Extension: we extend $g = e \circ q$ to a larger domain X

Product of X and Y is an object $P \cong X \times Y$ with two arrows $pr_X : P \to X$ and $pr_Y : P \to Y$ with a universal property: if $f \in Ar(Z, X)$ and $g \in Ar(Z, Y)$ then there is a unique arrow $h : Z \to P$ with $f = pr_X \circ h$ and $g = pr_Y \circ h$

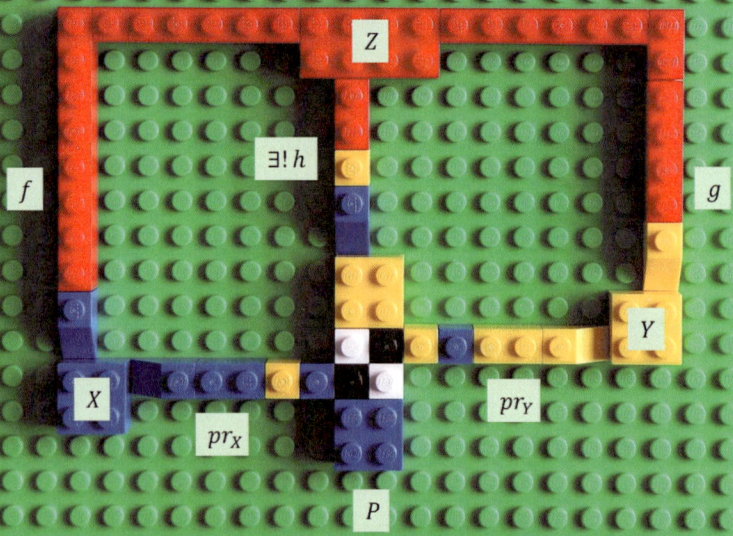

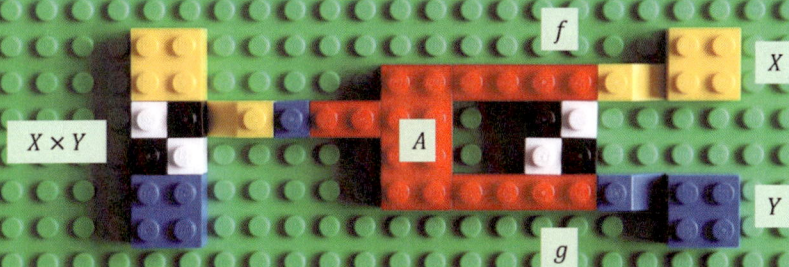

Product of arrows: $Ar(A, X \times Y) \cong Ar(A, X) \times Ar(A, Y)$

$X \times Y$

A

f

g

X

Y

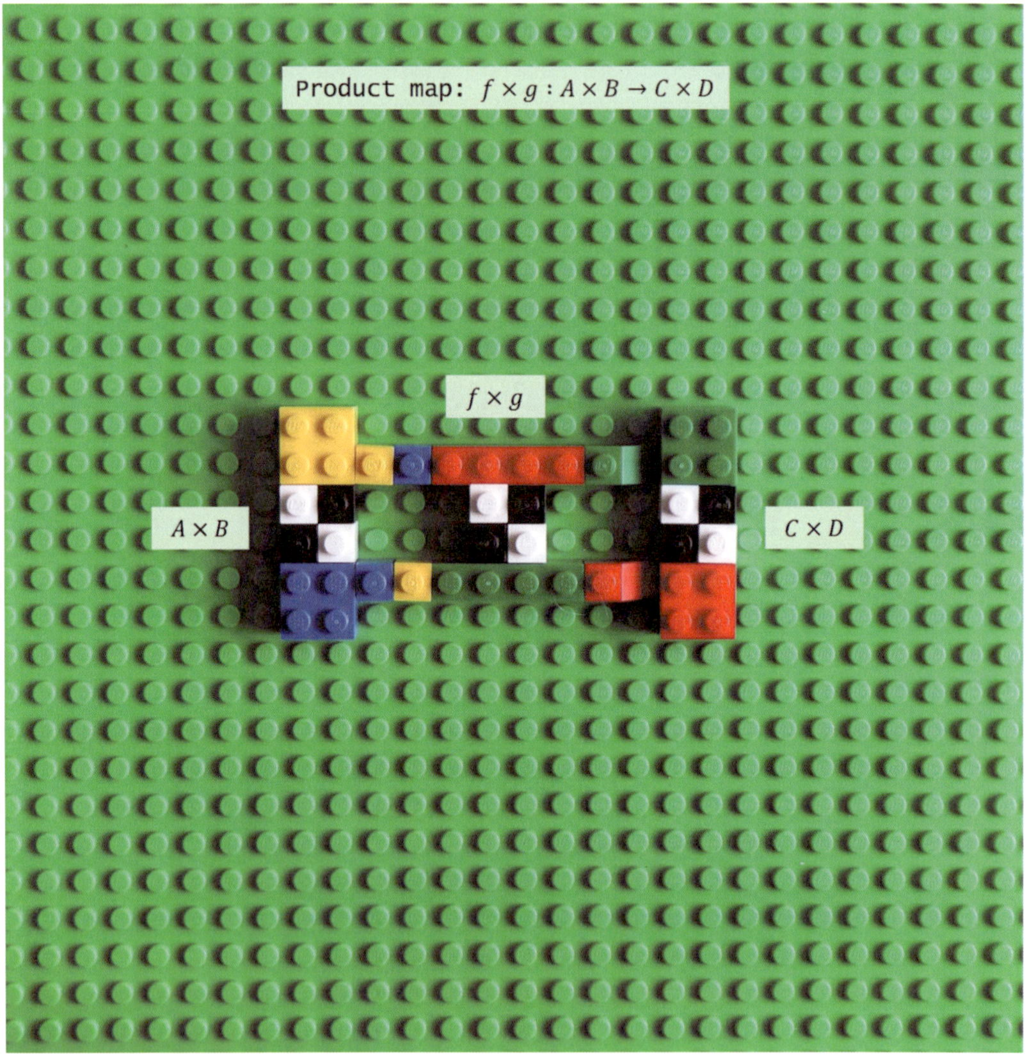

Product map: $f \times g : A \times B \to C \times D$

$f \times g$

$A \times B$

$C \times D$

Coproduct (sum) of X and Y is an object $S \cong X \sqcup Y$ with two arrows $i_X : X \to S$ and $i_Y : Y \to S$ with a universal property:
if $f \in Ar(X, Z)$ and $g \in Ar(Y, Z)$ then there is a unique arrow $q : S \to Z$ with $f = q \circ i_X$ and $g = q \circ i_Y$

Z

f $\exists! q$ g

Y

X i_Y

i_X

S

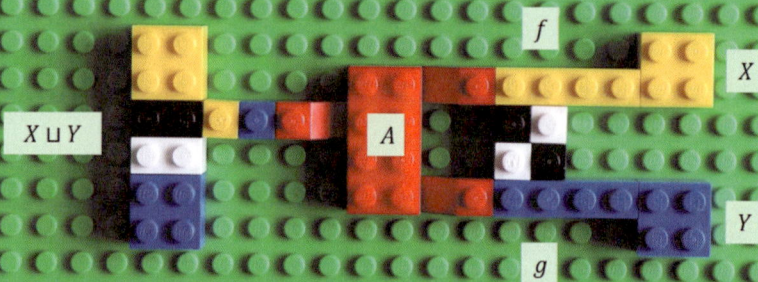

Product of arrows: $Ar(X \sqcup Y, A) \cong Ar(X, A) \times Ar(Y, A)$

Coproduct map: $f \sqcup g : A \sqcup B \to C \sqcup D$

$f \sqcup g$

$A \sqcup B$

$C \sqcup D$

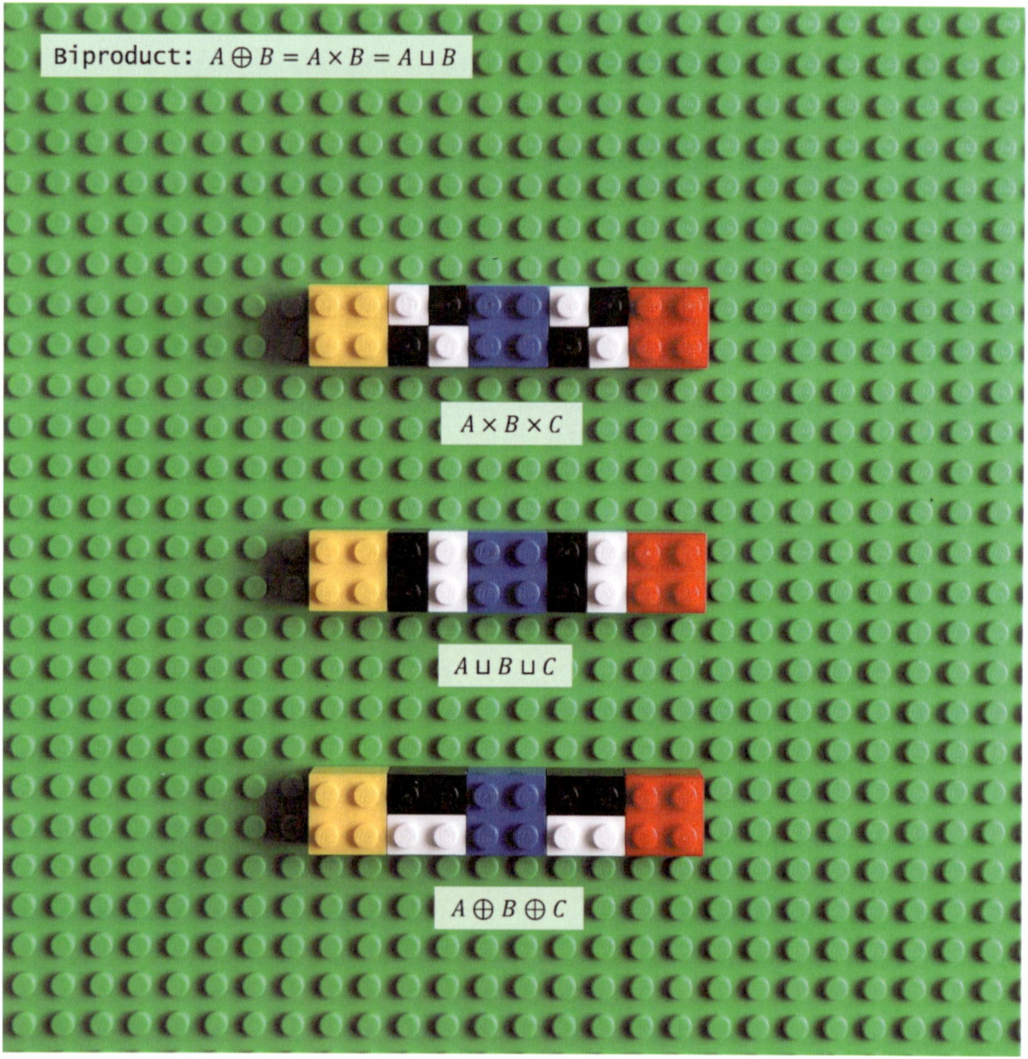

Biproduct: $A \oplus B = A \times B = A \sqcup B$

$A \times B \times C$

$A \sqcup B \sqcup C$

$A \oplus B \oplus C$

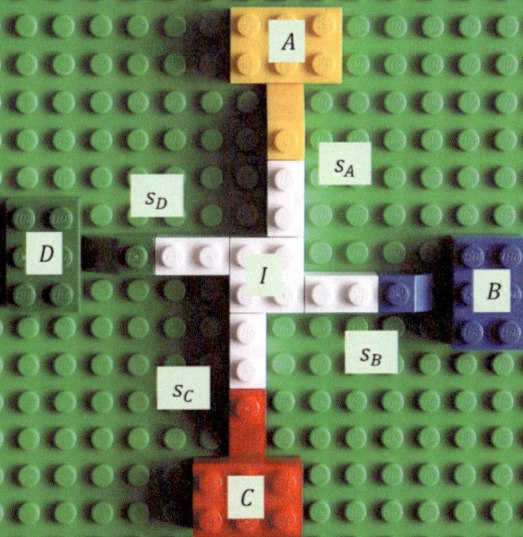

Initial object $I \in Ob(\mathbf{C})$: $\forall X \in Ob(\mathbf{C}) \; \exists ! \, s_X \in Ar(I, X)$

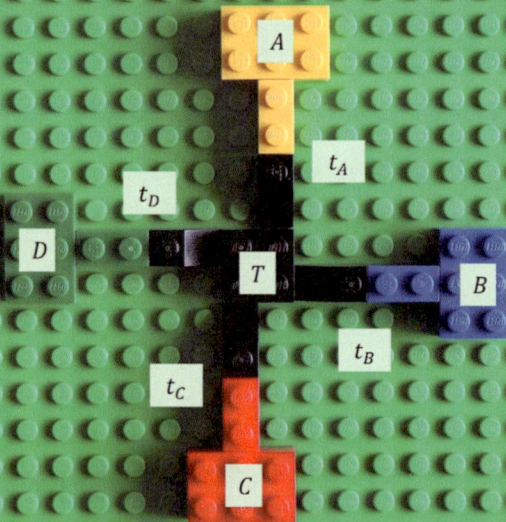

Terminal object $T \in Ob(\pmb{C})$: $\forall X \in Ob(\pmb{C}) \; \exists! \, t_X \in Ar(X,T)$

A

t_A

t_D

D

T

B

t_B

t_C

C

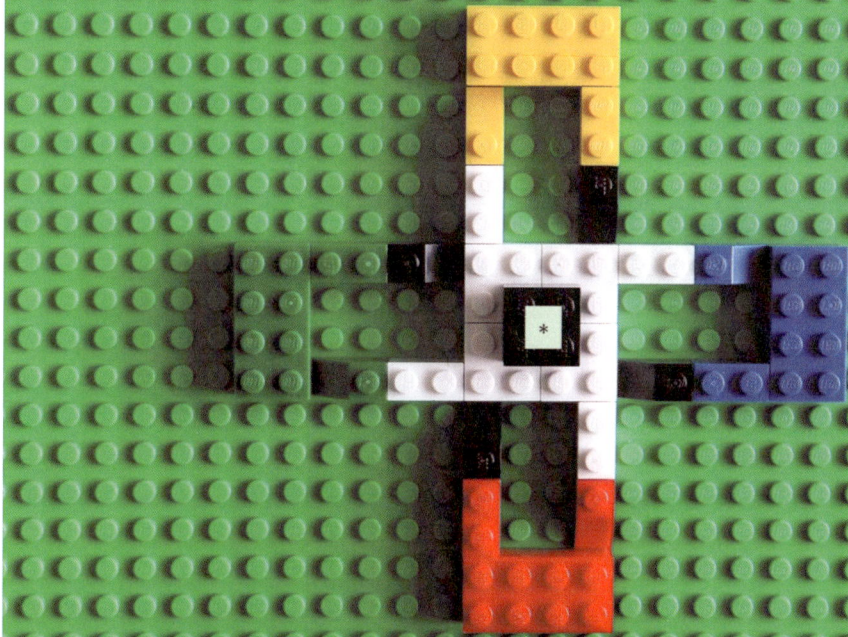

Pointed category C_*.
$* \in Ob(C_*)$ is initial and terminal object

Category of pointed sets **Set**$_*$
$X, Y \in Ob(\textbf{Set}_*)$, $x_o \in X$ and $y_o \in Y$ - basepoints
$f \in Ar(X, Y)$ - morphism from (X, x_o) to (Y, y_o) with $f(x_o) = y_o$

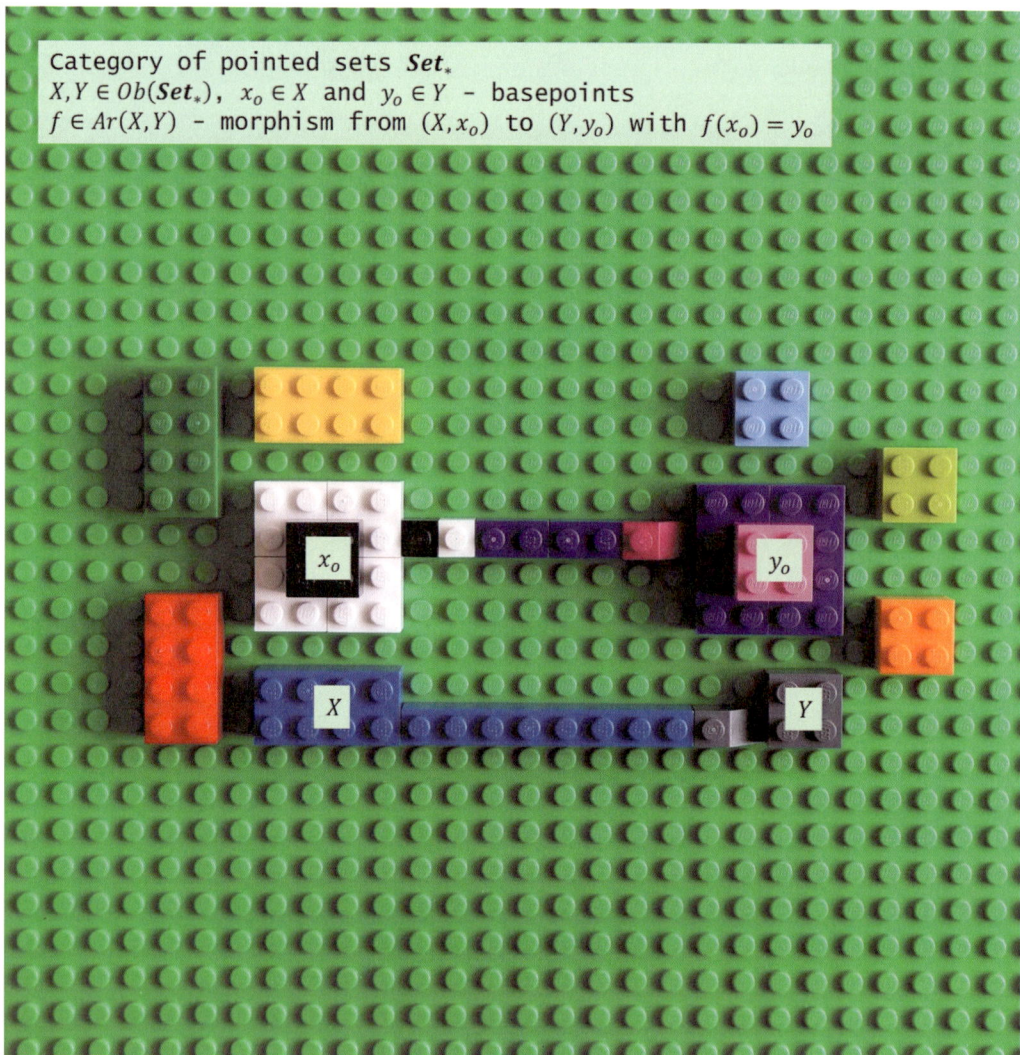

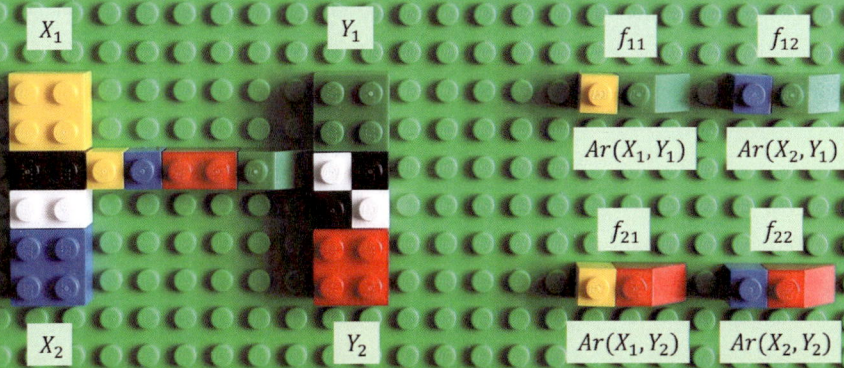

Matrix representation of arrows

$\forall f \in Ar(X_1 \sqcup X_2, Y_1 \times Y_2)$

$f \cong \begin{pmatrix} f_{11} & f_{12} \\ f_{21} & f_{22} \end{pmatrix} \quad f_{ij} \in Ar(X_j, Y_i)$

X_1

Y_1

f_{11}

f_{12}

$Ar(X_1, Y_1)$

$Ar(X_2, Y_1)$

f_{21}

f_{22}

$Ar(X_1, Y_2)$

$Ar(X_2, Y_2)$

X_2

Y_2

Monoid operation $M \times M \to M$

Visual Category Theory

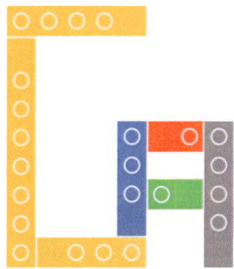

Dmitry Vostokov

brick by brick

$Part_3 \in C^{Parts}$

Visual Category Theory Brick by Brick, Part 3: Using LEGO® to Teach Abstract Mathematics

Published by OpenTask, Republic of Ireland

OpenTask books and magazines are available through booksellers and distributors worldwide. For further information or comments, send requests to press@opentask.com.

A CIP catalog record for this book is available from the British Library.

ISBN-13: 978-1912636426 (Paperback)

Revision 1.00 (April 2020)

Consider covariant functors $L: D \to C$ and $R: C \to D$
between categories C and D, objects $X \in D$ and $Y \in C$,
and the following arrows in categories C and D:
$Ar(C) \ni \alpha: L(X) \to Y$ and $Ar(D) \ni \alpha': X \to R(Y)$

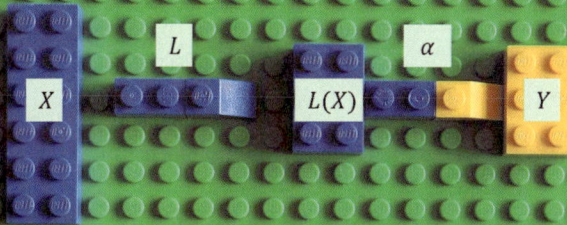

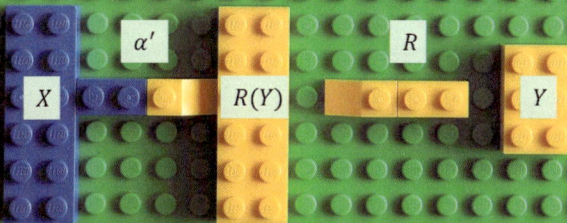

In addition to 3 books mentioned in Part$_1$, we also used the following references:

- Category Theory by Steve Awodey
- Topoi: The Categorical Analysis of Logic by Robert Goldblatt
- The Theory of Mathematical Structures by Jiří Adámek
- Mathematics of the Transcendental by Alain Badiou
- Memory Evolutive Systems: Hierarchy, Emergence, Cognition by Andrée Ehresmann and Jean-Paul Vanbremeersch

These functors $L: \boldsymbol{D} \to \boldsymbol{C}$ and $R: \boldsymbol{C} \to \boldsymbol{D}$ between categories $\boldsymbol{C}$ and $\boldsymbol{D}$ are called adjoint to each other (left adjoint and right adjoint respectively) or adjunctions if there is an equivalence between arrows $Ar(\boldsymbol{C}) \ni \alpha: L(X) \to Y$ and $Ar(\boldsymbol{D}) \ni \alpha': X \to R(Y)$ that respect morphisms between sources and morphisms between targets:

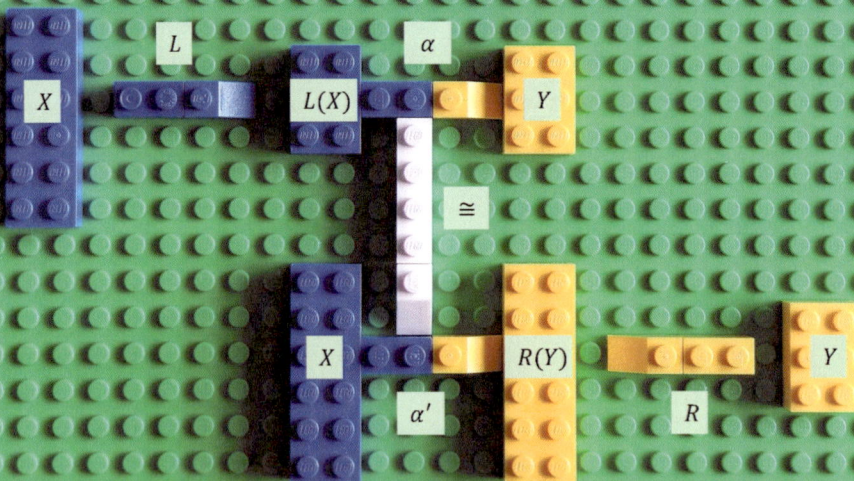

If functors $L: \boldsymbol{D} \to \boldsymbol{C}$ and $R: \boldsymbol{C} \to \boldsymbol{D}$ between categories $\boldsymbol{C}$ and $\boldsymbol{D}$ are adjoint then the following outer and inner diagrams commute ($X, U \in Ob(\boldsymbol{D})$ and $Y, V \in Ob(\boldsymbol{C})$, $Ar(\boldsymbol{C}) \ni \alpha: L(X) \to Y$ and $Ar(\boldsymbol{D}) \ni \alpha': X \to R(Y)$, $Ar(\boldsymbol{C}) \ni \beta: L(U) \to V$ and $Ar(\boldsymbol{D}) \ni \beta': U \to R(V)$)

A diagram D with a shape I is a covariant functor $D:I \to C$.
The collection of all diagrams in a category C with the
given shape I is a diagram category C^I.

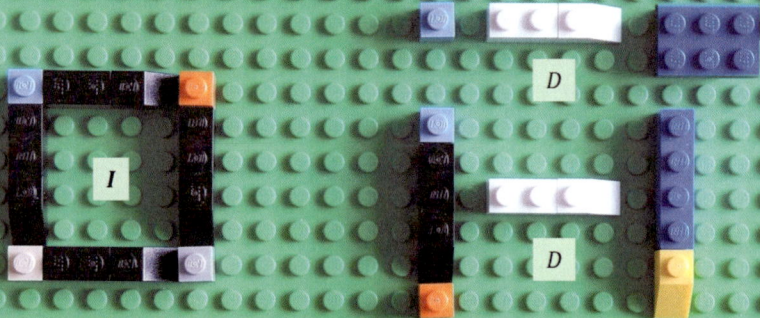

I D D

$D, D' \in C^I$

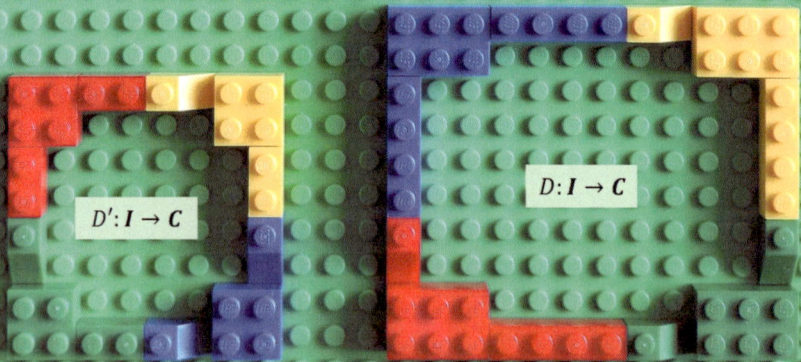

$D':I \to C$ $D:I \to C$

A cone $(C^{\rightarrow}, c_i)$ to a diagram $D: \mathbf{I} \rightarrow \mathbf{C}$ is $C^{\rightarrow} \in Ob(\mathbf{C})$ and a family of arrows $Ar(\mathbf{C}) \ni c_i : C^{\rightarrow} \rightarrow D(i)$, $\forall i \in \mathbf{I}$. It can be interpreted as the observability of a diagram or linking to all objects.

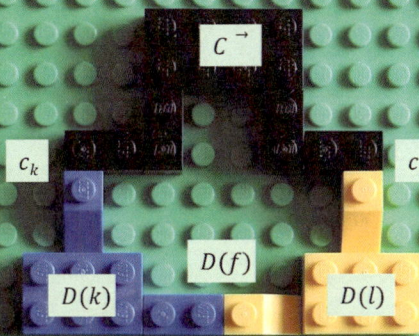

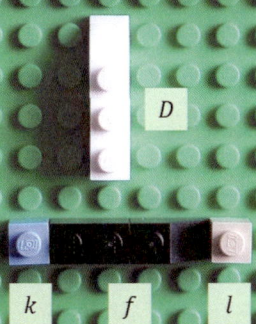

$C^{\rightarrow}$

c_k

c_l

$D(f)$

$D(k)$

$D(l)$

D

k f l

A cocone $(C^{\leftarrow}, c_i)$ to a diagram $D: I \to C$ is $C^{\leftarrow} \in Ob(C)$ and a family of arrows $Ar(C) \ni c_i : D(i) \to C^{\leftarrow}$, $\forall i \in I$. It can be interpreted as a collective link (linking from all objects).

A diagram D may have different cones.
All these cones form a category of cones $\textbf{Cone}(D)$ to D.
There is also a category of cocones.
$(L, l_i), (X, x_i) \in \textbf{Cone}(D), \quad \forall i \in \textbf{I}$

A limit for a diagram is a cone that has a unique arrow to it from any other cone. It is also a terminal object in the category of cones.

$X, L^{\leftarrow} \in Ob(\mathbf{C})$, $\forall X \exists! f_X \in Ar(\mathbf{C}): X \to L^{\leftarrow}$

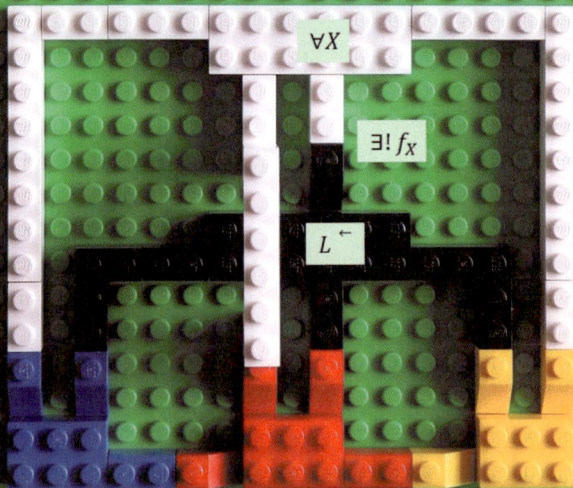

$\forall X$

$\exists! f_X$

$L^{\leftarrow}$

A colimit for a diagram is a cocone that has a unique
arrow from it to any other cocone. It is also an initial
object in the category of cocones.
$X, \vec{L} \in Ob(\mathbf{C}), \quad \forall X \exists! \, g_X \in Ar(\mathbf{C}): X \to \vec{L}$

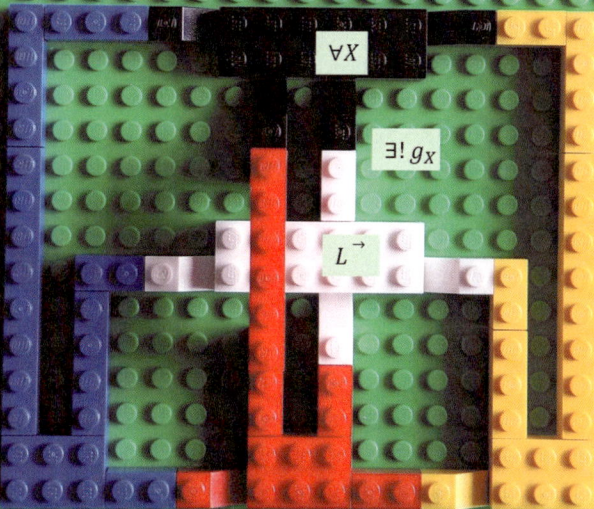

$\forall X$

$\exists! \, g_X$

$\vec{L}$

There are some common diagram types in category theory. One is
$X \to Z \leftarrow Y$. It may be completed by different solutions: $Q, S \in Ob(\mathbf{C})$

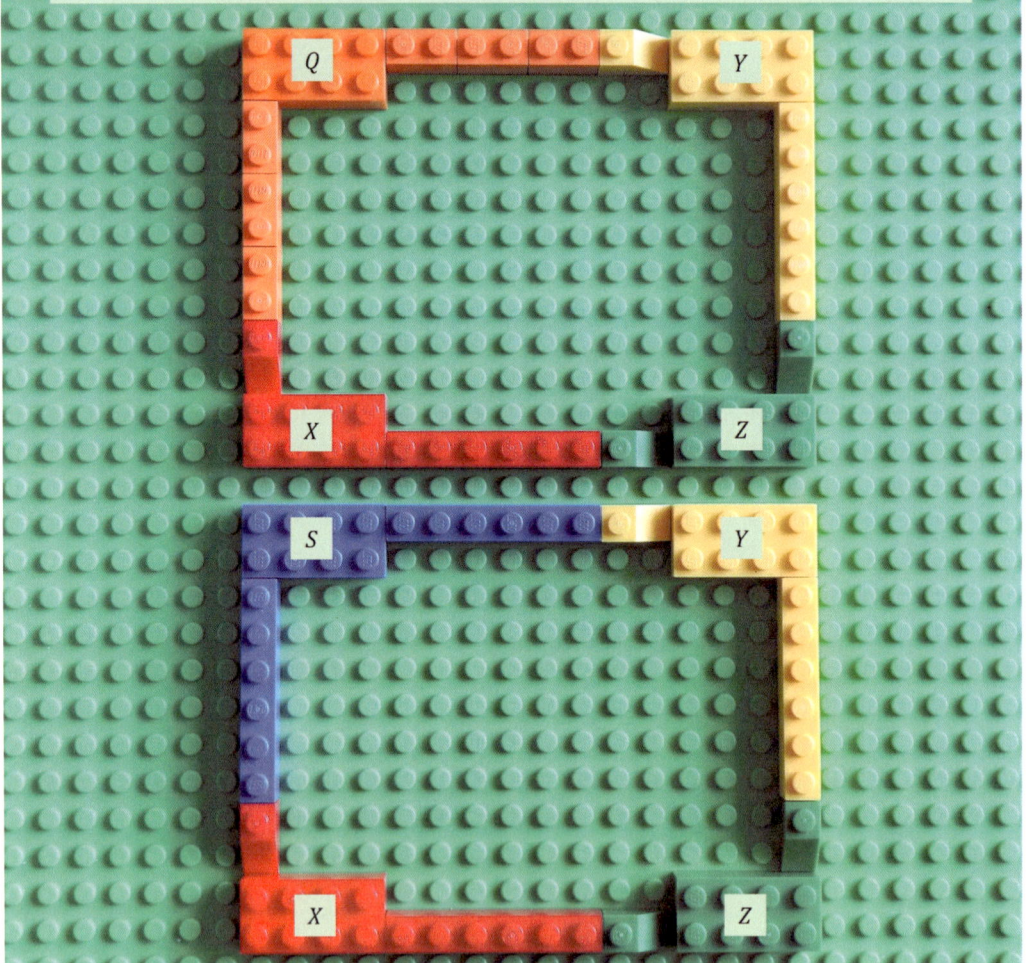

Q

Y

X

Z

S

Y

X

Z

A pullback is a limit of $X \to Z \leftarrow Y$ diagram (we omit some cone arrows):

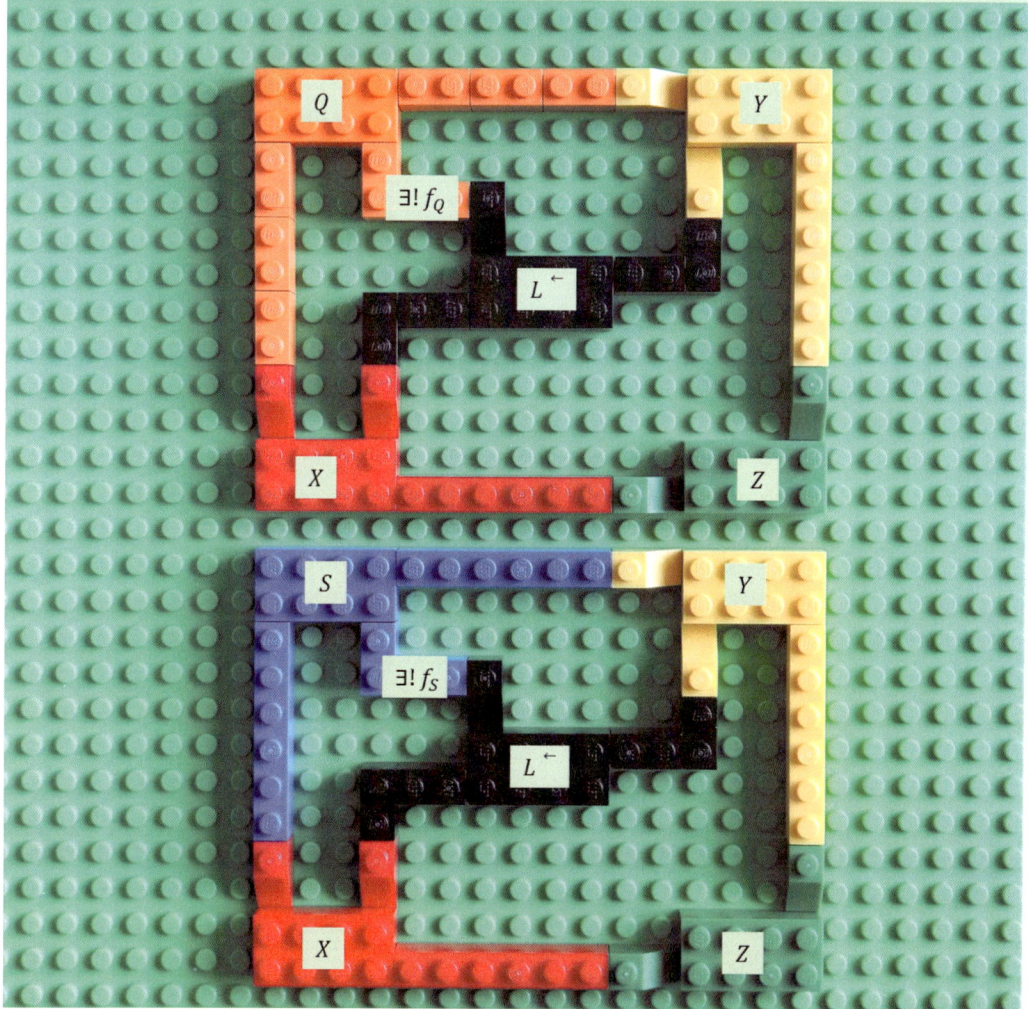

Another common diagram type in the category theory is $X \leftarrow Z \rightarrow Y$.
It may be completed by different solutions: $R, S \in Ob(\mathbf{C})$

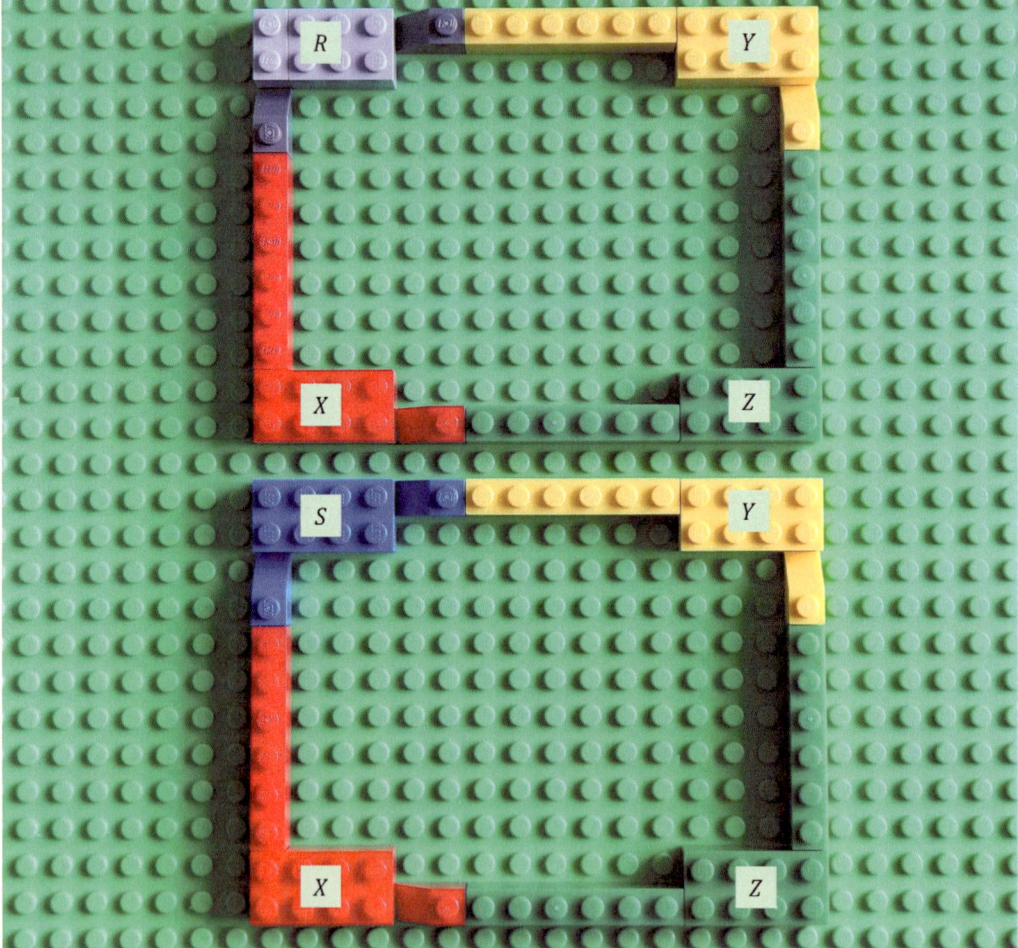

A pushout is a colimit of $X \leftarrow Z \rightarrow Y$ diagram (we omit some cocone arrows):

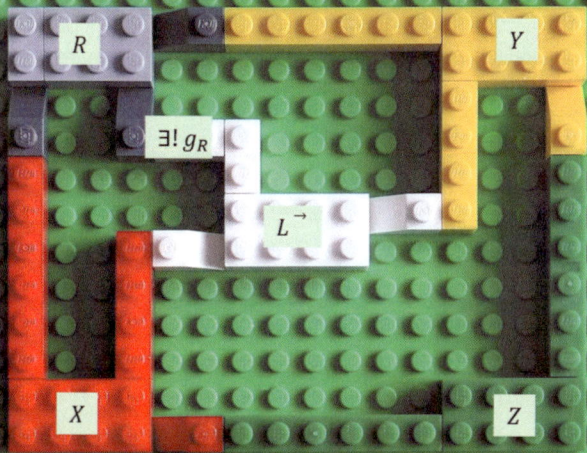

A pullback is equivalent to a product with a restriction:
$\{L^{\curvearrowleft}\} \cong \{(X,Y) \mid f(X) = g(Y) = Z\}$

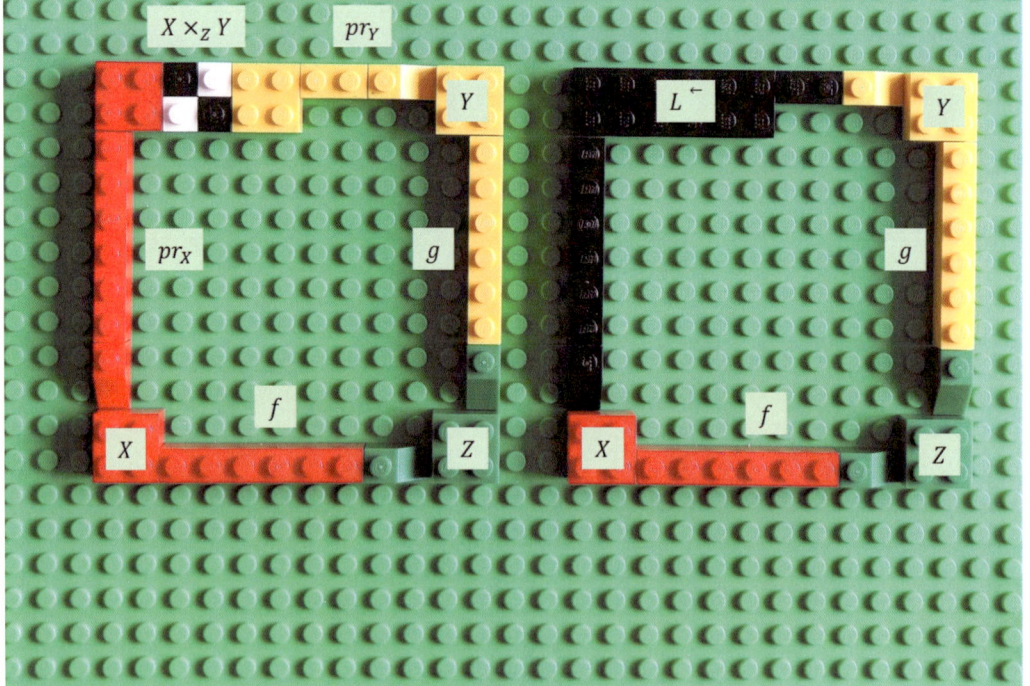

$X \times_Z Y$

pr_Y

Y

pr_X

g

f

X

Z

$L^{\leftarrow}$

Y

g

f

X

Z

A pushout is equivalent to a coproduct with a restriction: $\{L \overset{\rightarrow}{\ } \} \cong \{X + Y \mid X \sim Y \text{ if } \exists Z : h(Z) = X, k(Z) = Y\}$ or $X \cap Y$ in the set interpretation

$X \sqcup_Z Y$

Y

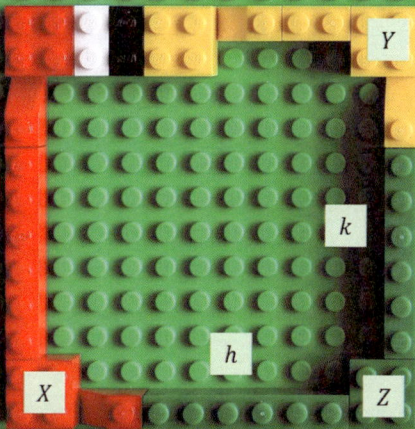

$L \overset{\rightarrow}{\ }$

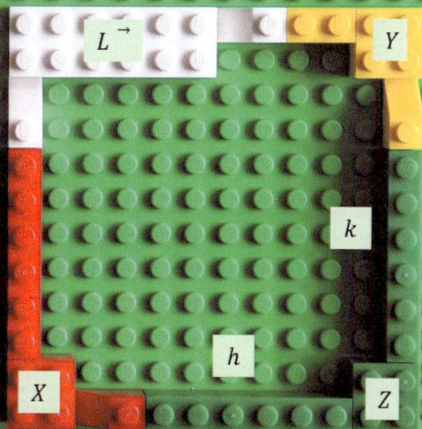

Y

k

k

h

h

X

Z

X

Z

[This page intentionally left blank]

Visual Category Theory

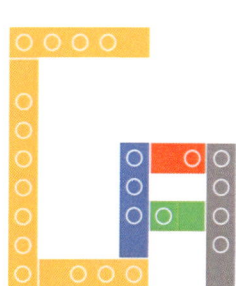

Dmitry
Vostokov

brick by brick

$$Part_4 \in C^{Parts}$$

Visual Category Theory Brick by Brick, Part 4: Using LEGO® to Teach Abstract Mathematics

Published by OpenTask, Republic of Ireland

OpenTask books and magazines are available through booksellers and distributors worldwide. For further information or comments, send requests to press@opentask.com.

A CIP catalog record for this book is available from the British Library.

ISBN-13: 978-1912636433 (Paperback)

Revision 1.01 (May 2020)

Categories may not have objects as sets or arrows as functions. For example, in the category C below, arrows are 2x2 matrices of numbers with arrow composition similar to matrix multiplication.

Numbers

$f \in Ar(C) : A \to B$
$A, B \in Ob(C)$
$|A|, |B| = 2$
A, B are any sets
with cardinality 2

$f \in Ar(C)$ $\qquad$ $g \in Ar(C)$ $\qquad$ $g \circ f \in Ar(C)$

In addition to 8 books mentioned in Part$_1$ and Part$_3$, we also used the following references:

- From Categories to Homotopy Theory by Birgit Richter
- From a Geometrical Point of View: A Study of the History and Philosophy of Category Theory by Jean-Pierre Marquis
- Machine Learning Brick by Brick, Epoch 1 by Dmitry Vostokov (algebraic notation for matrix multiplication)

In Part 2, we introduced a monoid operation between different objects. A monoid category M has only one object but many arrows with their composition as a monoid operation.

$\{M\} = Ob(M)$ $m_0 = id_M$

$m_1 \in Ar(M) : M \to M$

$m_2 \in Ar(M) : M \to M$

$m_i \in Ar(M) : M \to M$

m_1 m_1 $m_2 = m_1 \circ m_1$

A monoid object $M \in Ob(\boldsymbol{C})$ has $m \in Ar(\boldsymbol{C}) : M \times M \to M$ operation.
A group object $G \in Ob(\boldsymbol{C})$ is a monoid object with an inverse
operation $inv \in Ar(\boldsymbol{C}) : G \to G$.

$g \in Ar(\boldsymbol{C}) : G \times G \to G$
$A, AA, A^{-1} \in G$ (Note: these are not objects of $\boldsymbol{C}$, g and inv
act internally on G which has a group structure)

A group category G is a monoid category (only one object but many arrows) with each arrow having an inverse (equivalence).

$\{G\} = Ob(\mathbf{G})$

$\forall g \in Ar(\mathbf{G}) : G \to G \; \exists! \, g^{-1} \in Ar(\mathbf{G}) : G \to G$ with $g \circ g^{-1} = g^{-1} \circ g = id_G$

g_1

g_1^{-1}

g_1 g_1 $g_2 = g_1 \circ g_1$

g_1^{-1} g_1^{-1} $g_2^{-1} = g_1^{-1} \circ g_1^{-1}$

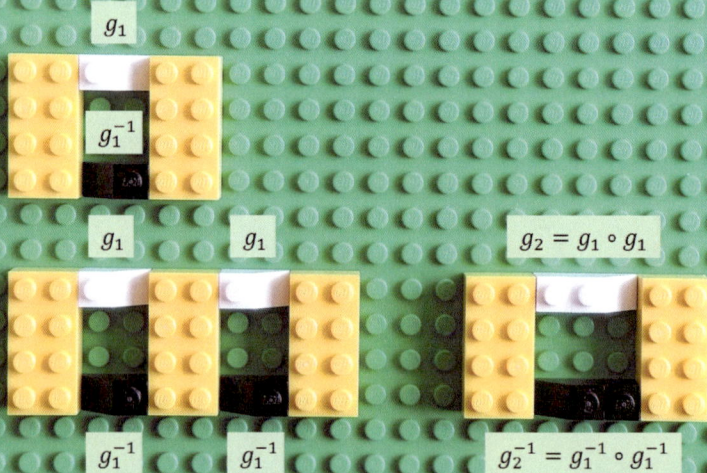

An opposite or dual category C^{op} is a category with the same objects of category C but reversed arrows.

$$\forall f \in Ar(C^{op}) : X \to Y \; \exists f' \in Ar(C) : Y \to X$$

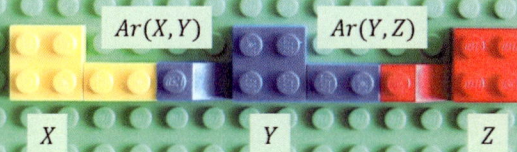

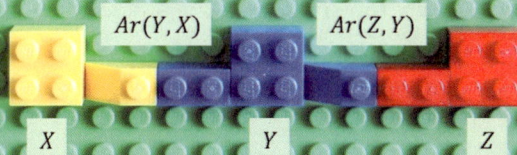

An arrow category C^{ar} is a category with the same arrows of category C as objects.

$Ob(C)$

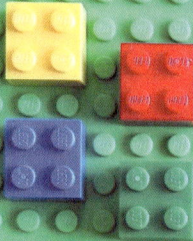

$Ar(C)$

$Ob(C^{ar})$

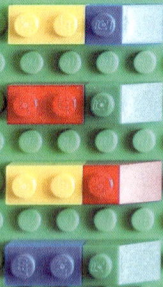

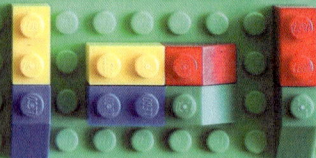

$(g_1, g_2) \in Ar(C^{ar})$

There are two functors from an arrow category C^{ar} to category C.

Dom

Codom

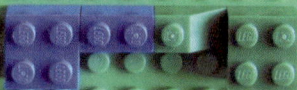

A slice category C/C is a category with the following properties:

$Ob(C/C) = \{f \in Ar(C) | codom(f) = C\}$
$Ar(C/C) = \{Ar(Ar(X,C), Ar(X',C)) | \forall f \in Ar(X,C), f' \in Ar(X',C) \, \exists g \in Ar(X,X'), f' \circ g = f\}$

$Ob(C)$ $Ar(C)$

C

X'

X

g

$Ob(C/C)$

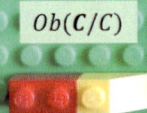

f f'

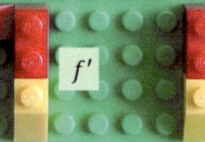

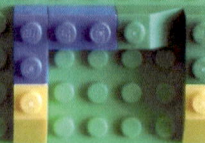

There is a functor from a slice category $\mathcal{C}/C$ to category $\mathcal{C}$ without C.

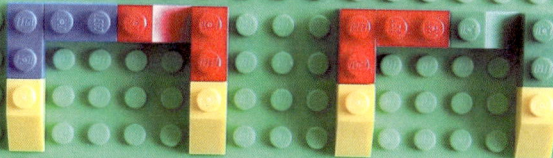

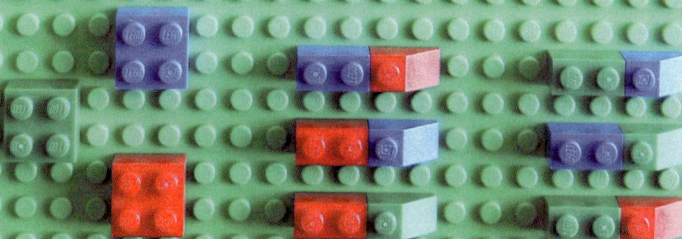

A coslice category C/C is a category with the following properties:

$Ob(C/C) = \{f \in Ar(C)|dom(f) = C\}$
$Ar(C/C) = \{Ar(Ar(C,X), Ar(C,X'))|\forall f \in Ar(C,X), f' \in Ar(C,X') \exists h \in Ar(X,X'), f' = h \circ f\}$

$Ob(C)$

$Ar(C)$

$Ob(C/C)$

f h f'

 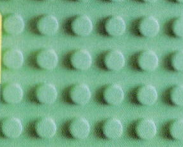

A forgetful functor F from category C to category D drops some arrows from C (less structure in D).

Ob(C)

Ar(C)

F

Ob(D)

Ar(D)

An arrow $f \in Ar(\mathbf{C}) : X \to Y$ is a monomorphism if $\forall g, h \in Ar(\mathbf{C}) : M \to X$, $f \circ g = f \circ h \Rightarrow g = h$.

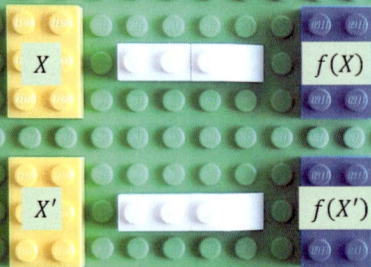

An injective function in **Sets**:
$f(X) = f(X') \Rightarrow X = X'$

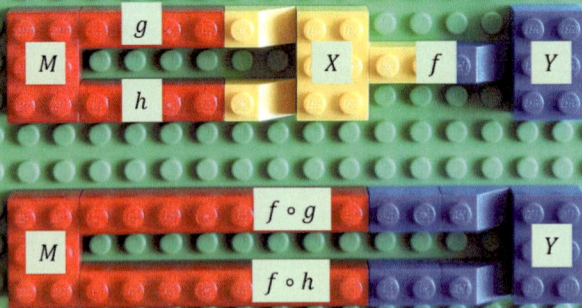

An arrow $f \in Ar(C) : X \to Y$ is an epimorphism if $\forall p, t \in Ar(C) : Y \to E$, $p \circ f = t \circ f \Rightarrow p = t$.

A surjective function in **Sets**:
$\forall Y \exists X f(X) = Y$

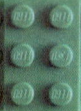

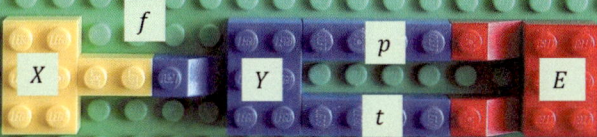

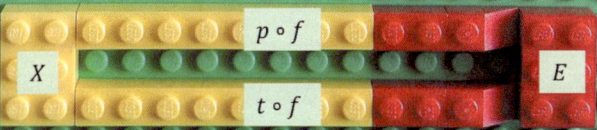

An arrow $f \in Ar(\boldsymbol{C}) : X \to Y$ is an isomorphism if it is both a monomorphism and epimorphism.

Visual Category Theory

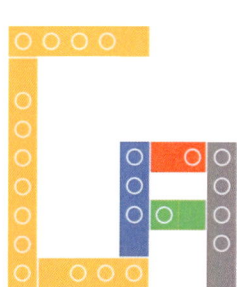

Dmitry Vostokov

Brick by Brick

Part$_5 \in C$ Parts

Visual Category Theory Brick by Brick, Part 5: Using LEGO® to Teach Abstract Mathematics

Published by OpenTask, Republic of Ireland

OpenTask books and magazines are available through booksellers and distributors worldwide. For further information or comments, send requests to press@opentask.com.

A CIP catalog record for this book is available from the British Library.

ISBN-13: 978-1912636440 (Paperback)

Revision 1.02 (October 2021)

Reading guide as a category C^{Parts} diagram.

$Part_1$

$Part_2$

$Part_3$

$Part_4$

$Part_5$

Let us consider a function of sets $f(z,x) : Z \times X \to Y$.

In addition to 11 books mentioned in Part$_1$, Part$_3$, and Part$_4$ we also used
Diagrammatic Immanence: Category Theory and Philosophy by Rocco Gangle

We fix the first variable $f(v,x) \in Y^X : X \to Y$.

By varying the parameter, we introduce a new (exponential) map to the function $f^E(z): Z \to Y^X$.

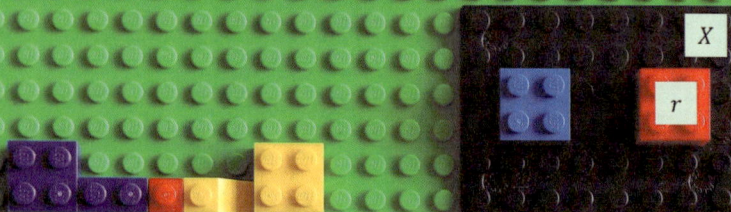

X

r

Y^X

Y

$$f^E(v)(r) = g(r) = f(v, r)$$
$$g = f^E(v) \in Y^X$$

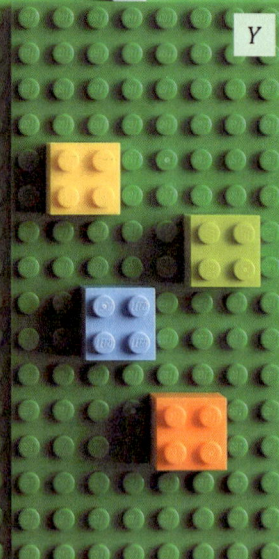

Z

$f^E(z)$

v

We also have an evaluation map that evaluates Y^X using X values: $e(g,x) : Y^X \times X \to Y$.

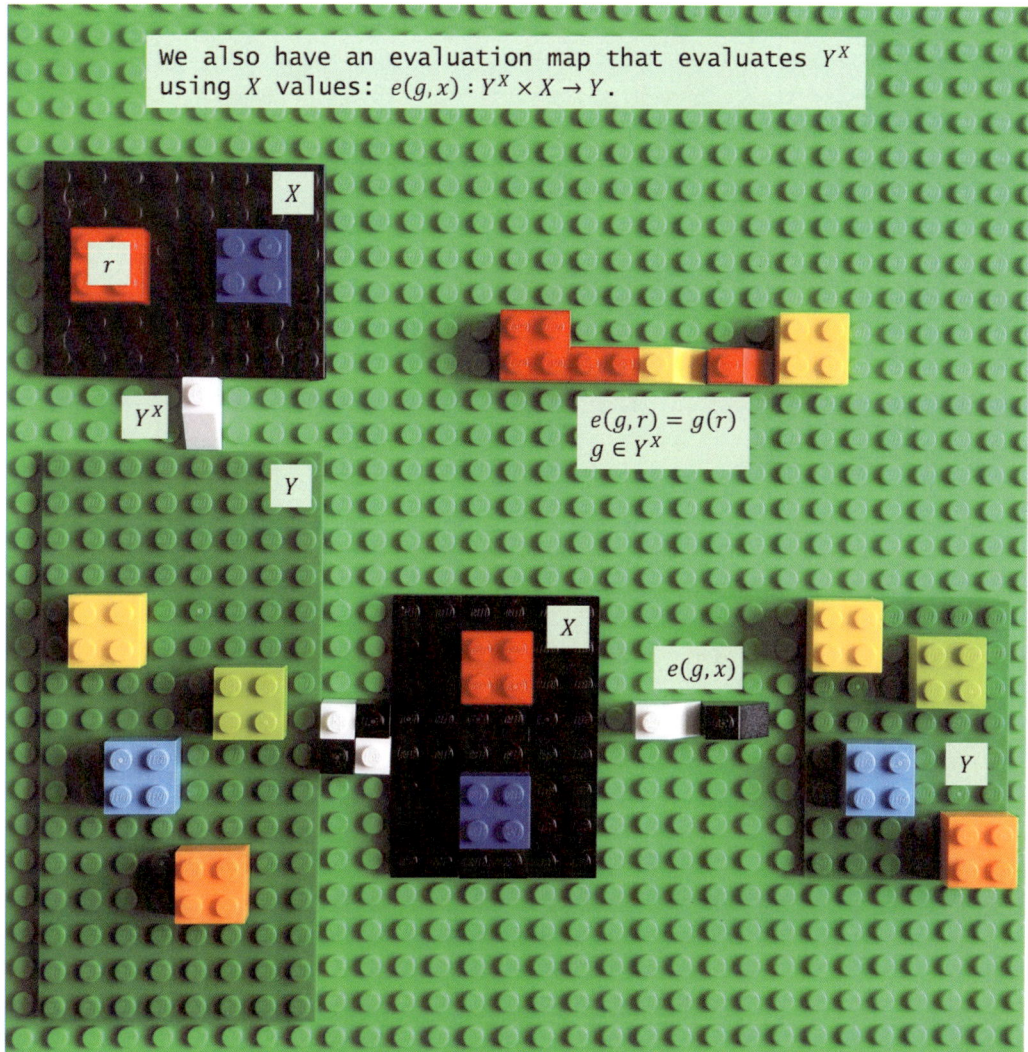

X

r

Y^X

Y

$e(g,r) = g(r)$
$g \in Y^X$

X

$e(g,x)$

Y

We can now combine exponential and evaluation functions into a commutative diagram.

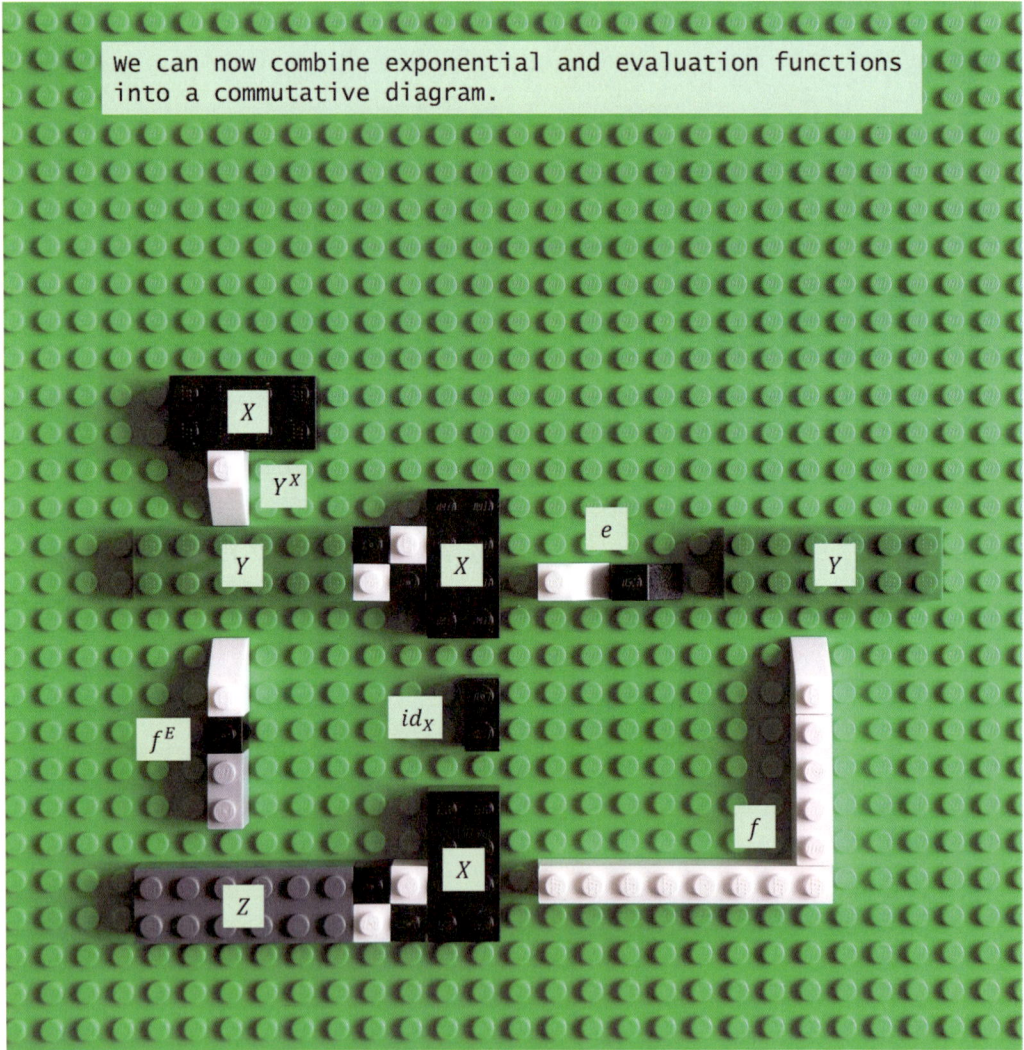

X

Y^X

Y X e Y

f^E id_X

Z X f

An exponential of objects $X, Y \in Ob(\boldsymbol{C})$ consists of:
$Y^X \in Ob(\boldsymbol{C})$
$e \in Ar(\boldsymbol{C}) : Y^X \times X \to Y$
$\forall Z \in Ob(\boldsymbol{C}) \; \forall f \in Ar(\boldsymbol{C}) : Z \times X \to Y \; \exists! f^E : Z \to Y^X \text{ and } e \circ (f^E \times id_X) = f$

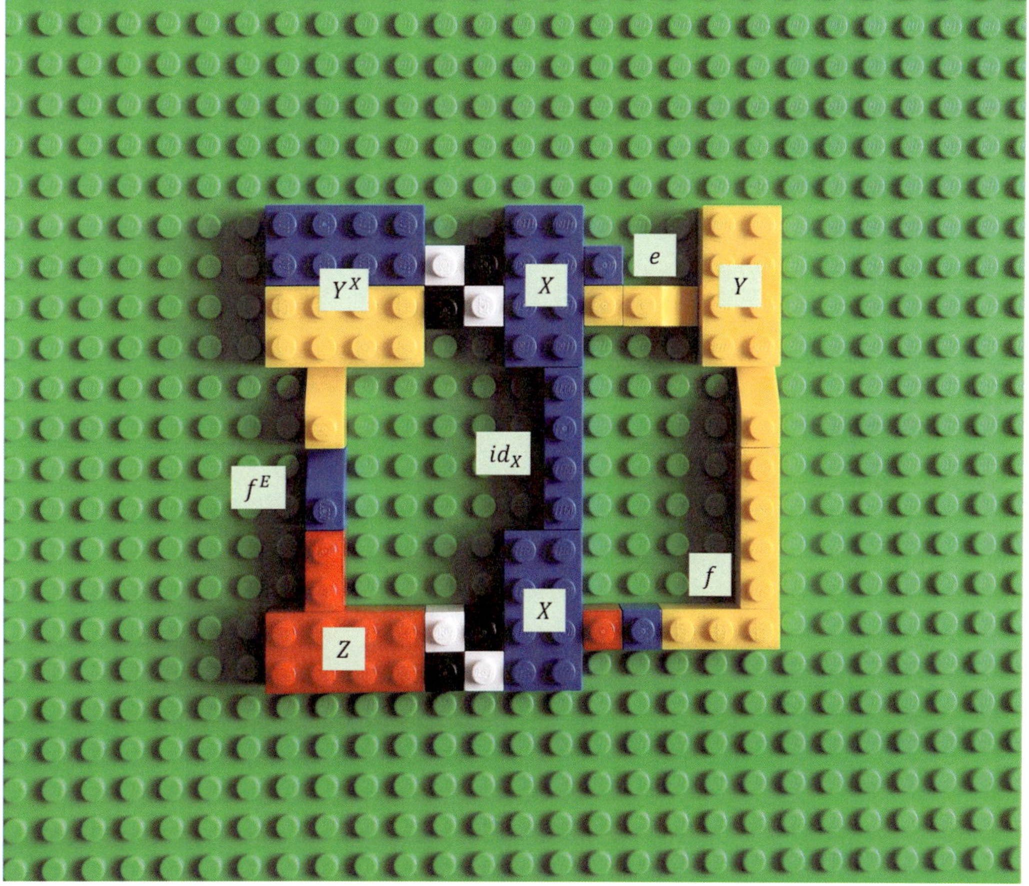

A category of subobjects of $X \in Ob(\mathbf{C})$ consists of objects: monomorphisms $m \in Ar(\mathbf{C}) : M \to X$ and arrows: $f \in Ar(\mathbf{C}/X): m \to m'$

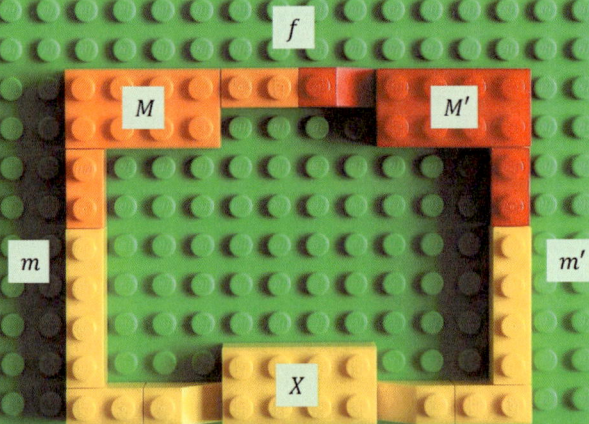

An equalizer of $f, g \in Ar(\mathbf{C}) : X \to Y$ consists of:
$E \in Ob(\mathbf{C})$
$e \in Ar(\mathbf{C}) : E \to X$ and $f \circ e = g \circ e$
$\forall Z \in Ob(\mathbf{C}), \forall z \in Ar(Z, X) \, \exists! u \in Ar(Z, E)$ and $z = e \circ u$

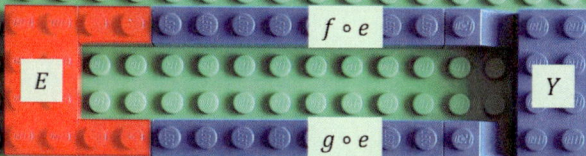

An equalizer is a monomorphism:
$\forall m, m' \in Ar(Z, E)\ e \circ m = e \circ m' \Rightarrow m = m'$.

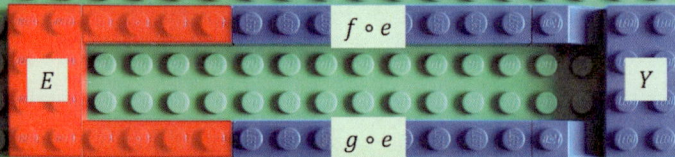

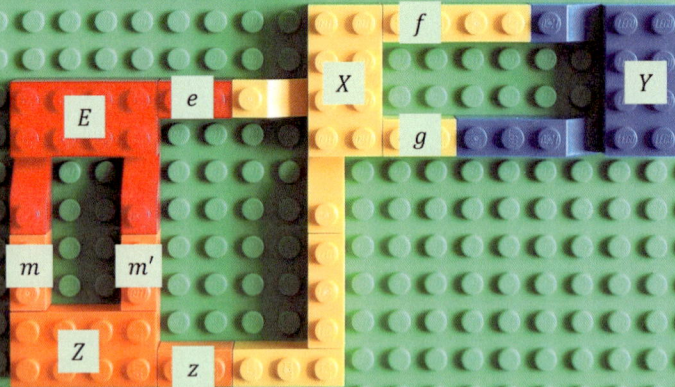

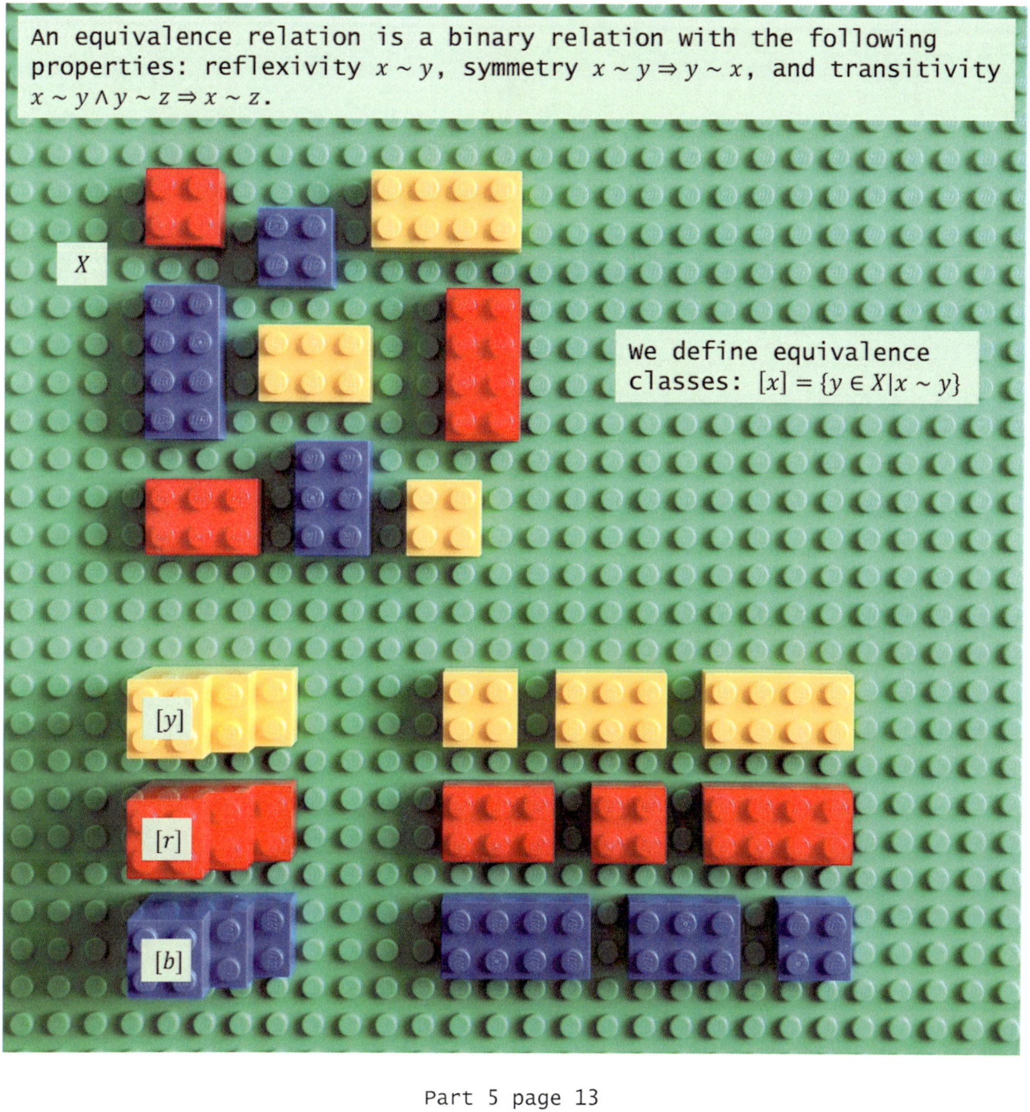

An equivalence relation is a binary relation with the following properties: reflexivity $x \sim y$, symmetry $x \sim y \Rightarrow y \sim x$, and transitivity $x \sim y \wedge y \sim z \Rightarrow x \sim z$.

X

We define equivalence classes: $[x] = \{y \in X | x \sim y\}$

$[y]$

$[r]$

$[b]$

A coequalizer of $f, g \in Ar(\mathbf{C}) : X \to Y$ consists of:
$C \in Ob(\mathbf{C})$
$c \in Ar(\mathbf{C}) : Y \to C$ and $c \circ f = c \circ g$
$\forall W \in Ob(\mathbf{C}), \forall w \in Ar(Y, W) \; \exists! u \in Ar(C, W)$ and $w = u \circ c$

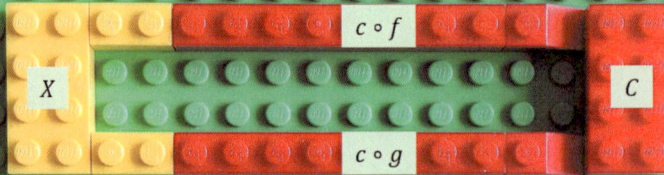

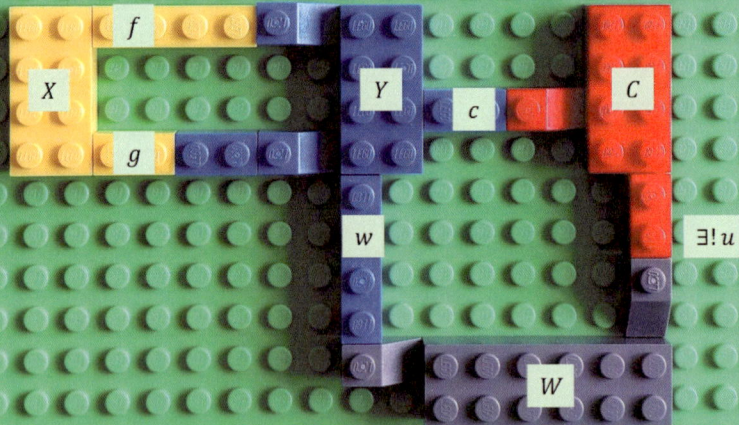

A coequalizer is an epimorphism:
$\forall p, p' \in Ar(C, W)\ p \circ c = p' \circ c \Rightarrow p = p'.$

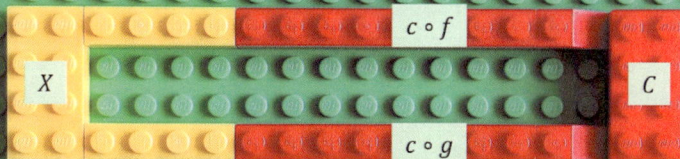

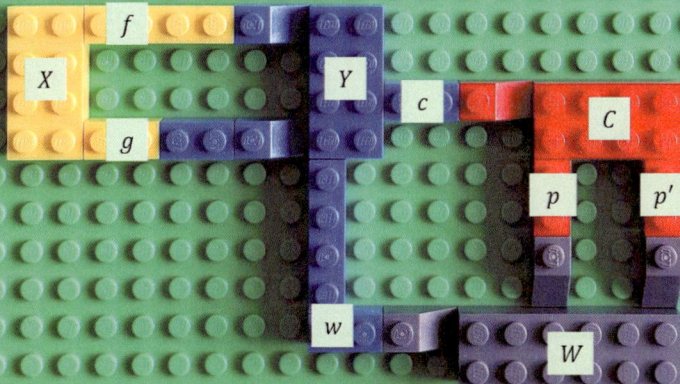

A congruence on C is an equivalence relation:
$f \sim g \Rightarrow Dom(f) = X = Dom(g) \wedge Cod(f) = Y = Cod(g)$.

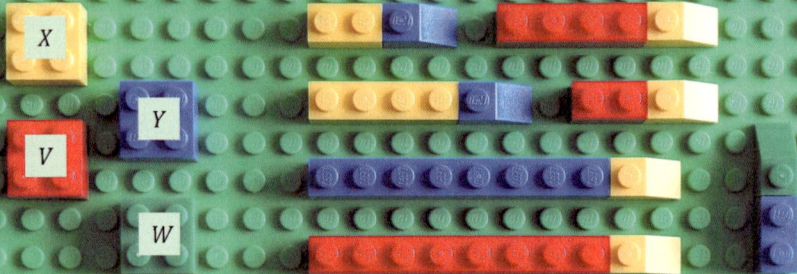

A congruence category $C^{\sim}$ is defined by:
$Ob(C^{\sim}) = Ob(C)$ and $Ar(C^{\sim}) = \{f \times g | g, f \in Ar(C), f \sim g\}$

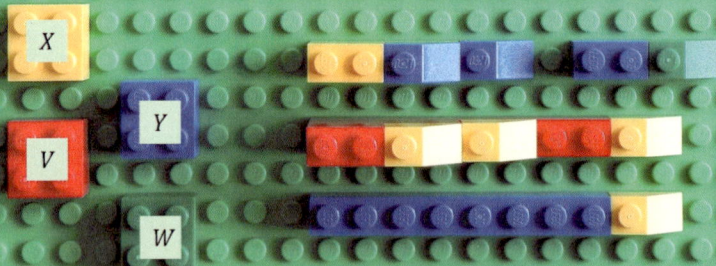

and $\forall v: V \to X \; \forall w: Y \to W \; f \sim g \Rightarrow w \circ f \circ v \sim w \circ g \circ v$.

A (covariant) morphism functor $F(X_0, -) : \boldsymbol{C} \to \boldsymbol{Sets}$
sends $X \in Ob(\boldsymbol{C})$ to $\{f | f \in Ar(X_0, X)\}$.

X_0

X

Y

$F(X_0, -)$

$\{f | f \in Ar(X_0, X)\}$

$\{g | g \in Ar(X_0, Y)\}$

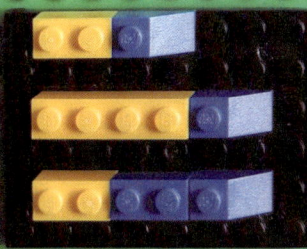

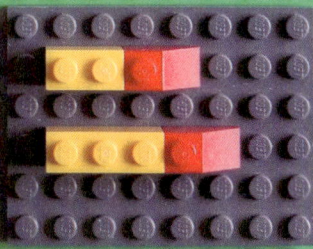

A presheaf is a contravariant functor $S : C \to \textbf{Sets}$ that sends objects to sets of object representations.

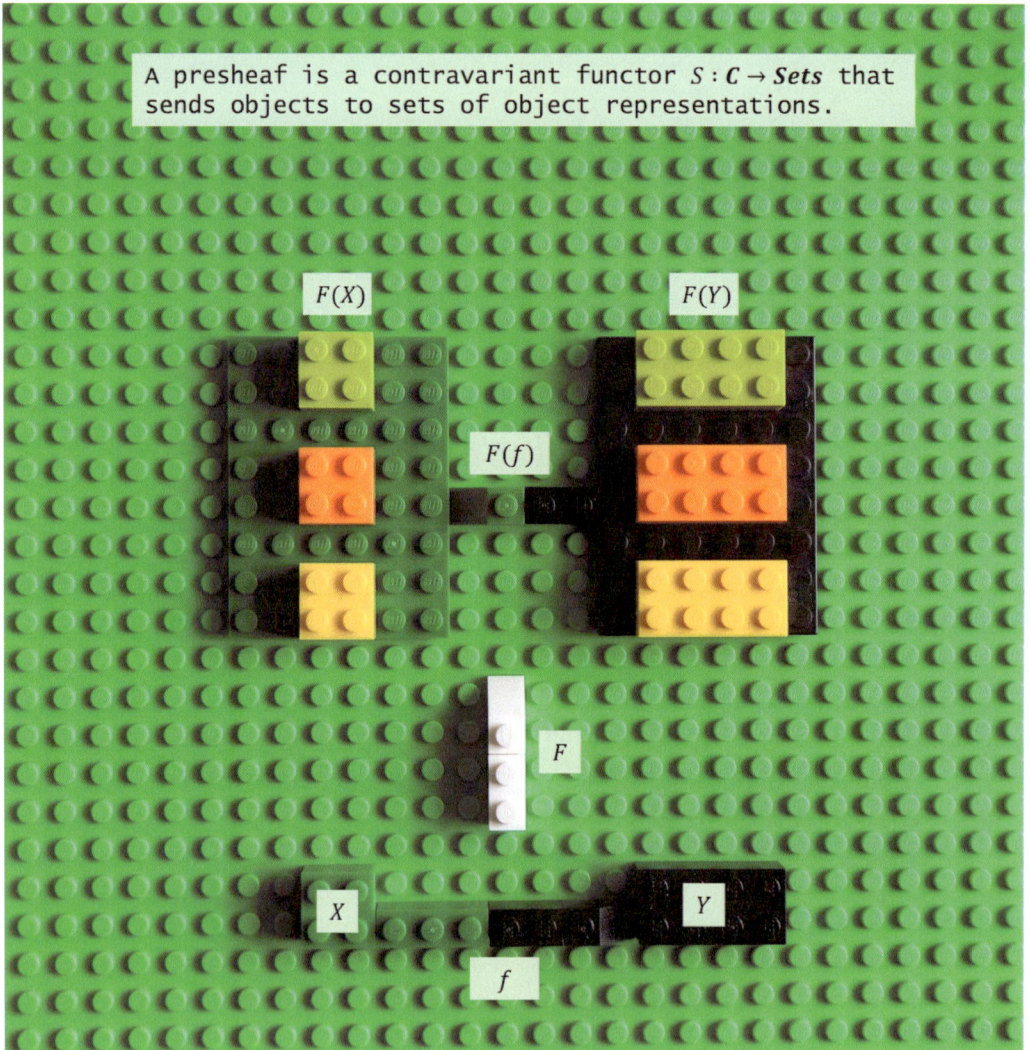

Visual Category Theory

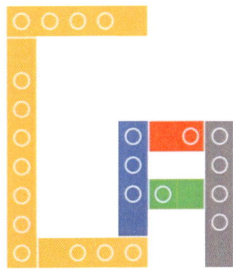

Dmitry Vostokov

rick by rick

$$Part_6 \in C^{Parts}$$

Visual Category Theory Brick by Brick, Part 6: Using LEGO® to Teach Abstract Mathematics

Published by OpenTask, Republic of Ireland

OpenTask books and magazines are available through booksellers and distributors worldwide. For further information or comments, send requests to press@opentask.com.

A CIP catalog record for this book is available from the British Library.

ISBN-13: 978-1912636457 (Paperback)

Revision 1.02 (June 2020)

In $Part_1$ we introduced natural transformations between functors. Natural transformations can be "vertically" composed.

$F, G, H : \mathbf{C} \to \mathbf{D}$, $X \in \mathbf{C}$, $F(X), G(X), H(X) \in \mathbf{D}$

$n_{FG} : F \to G$, $n_{FG}(X) : F(X) \to G(X)$, $n_{GH} : G \to H$, $n_{GH}(X) : G(X) \to H(X)$

$n_{FH} = n_{GH} \circ n_{FG} : F \to H$, $n_{FH}(X) = n_{GH}(X) \circ n_{FG}(X) : F(X) \to H(X)$

$\mathbf{C}$ X F $F(X)$ $\mathbf{D}$

n_{FG}

G $G(X)$

$H(X)$

n_{GH}

H

In addition to 12 books mentioned in $Part_1$, $Part_3$, $Part_4$, and $Part_5$ we also used:

· Iconicity and Abduction by Gianluca Caterina and Rocco Gangle
· Directed Algebraic Topology: Models of Non-Reversible Worlds by Marco Grandis

Natural transformations can also be "horizontally" composed (whisker composition, shown here as vertical for E and H functors due to square space constraints).

$E : A \to C$, $H : D \to B$, $F, G : C \to D$, $n_{FG}: F \to G$, $n_{FG}(X): F(X) \to G(X)$

$Hn_{FG}E = H \circ n_{FG} \circ E : HFE \to HGE$, $Hn_{FG}E(X) : HFE(X) \to HGE(X)$

$X \in A$, $E(X) \in C$, $FE(X), GE(X) \in D$, $HFE(X), HGE(X) \in B$

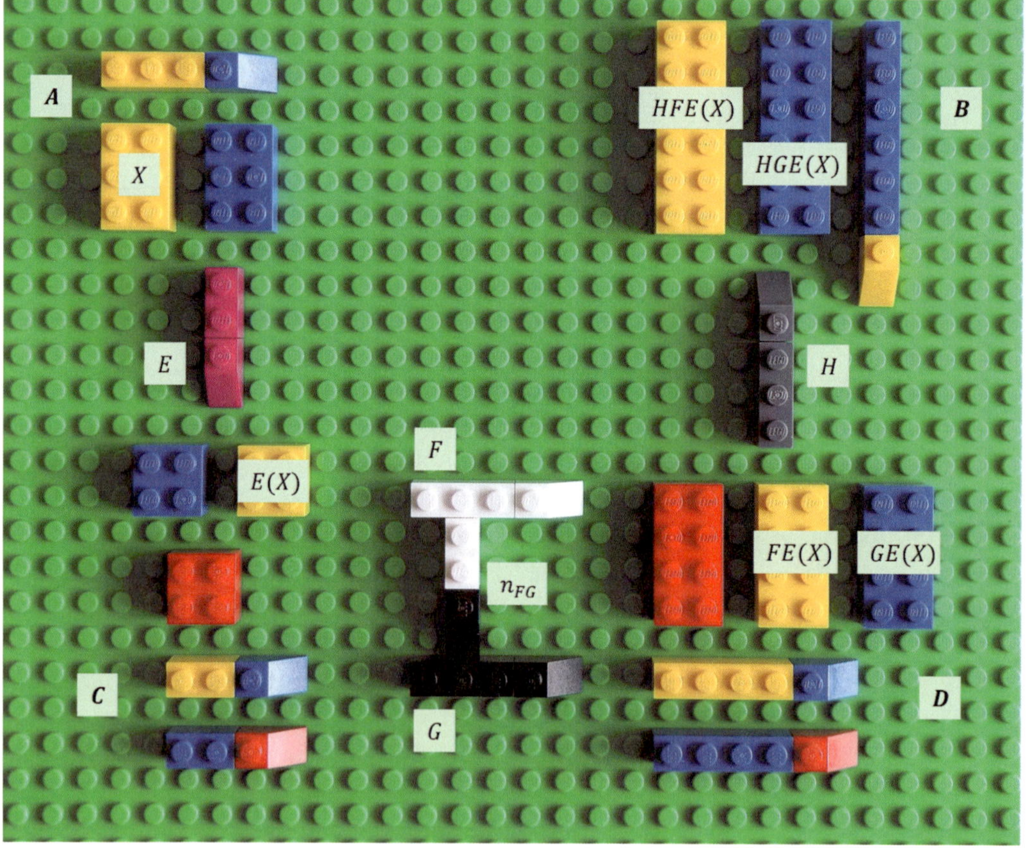

A

$HFE(X)$

$HGE(X)$

B

X

E

H

F

$E(X)$

n_{FG}

$FE(X)$

$GE(X)$

C

G

D

Part 6 page 4

The identity of a functor is a natural transformation:
$X \in \mathbf{C}, \ F(X) \in \mathbf{D}, \ id_F : F \to F, \ id_F(X) = id_{F(X)}$

$\mathbf{C}$

$\mathbf{D}$

$id_{F(X)}$

X

$F(X)$

F

id_F

F

An isomorphism of functors $F \cong G$ is an invertible natural transformation:

$F, G : \boldsymbol{C} \to \boldsymbol{D}$, $n_{FG} : F \to G$, $n_{GF} : G \to F$, $n_{GF} \circ n_{FG} = id_F$, $n_{FG} \circ n_{GF} = id_G$.

$\boldsymbol{C}$

$\boldsymbol{D}$

F

n_{FG} n_{GF}

G

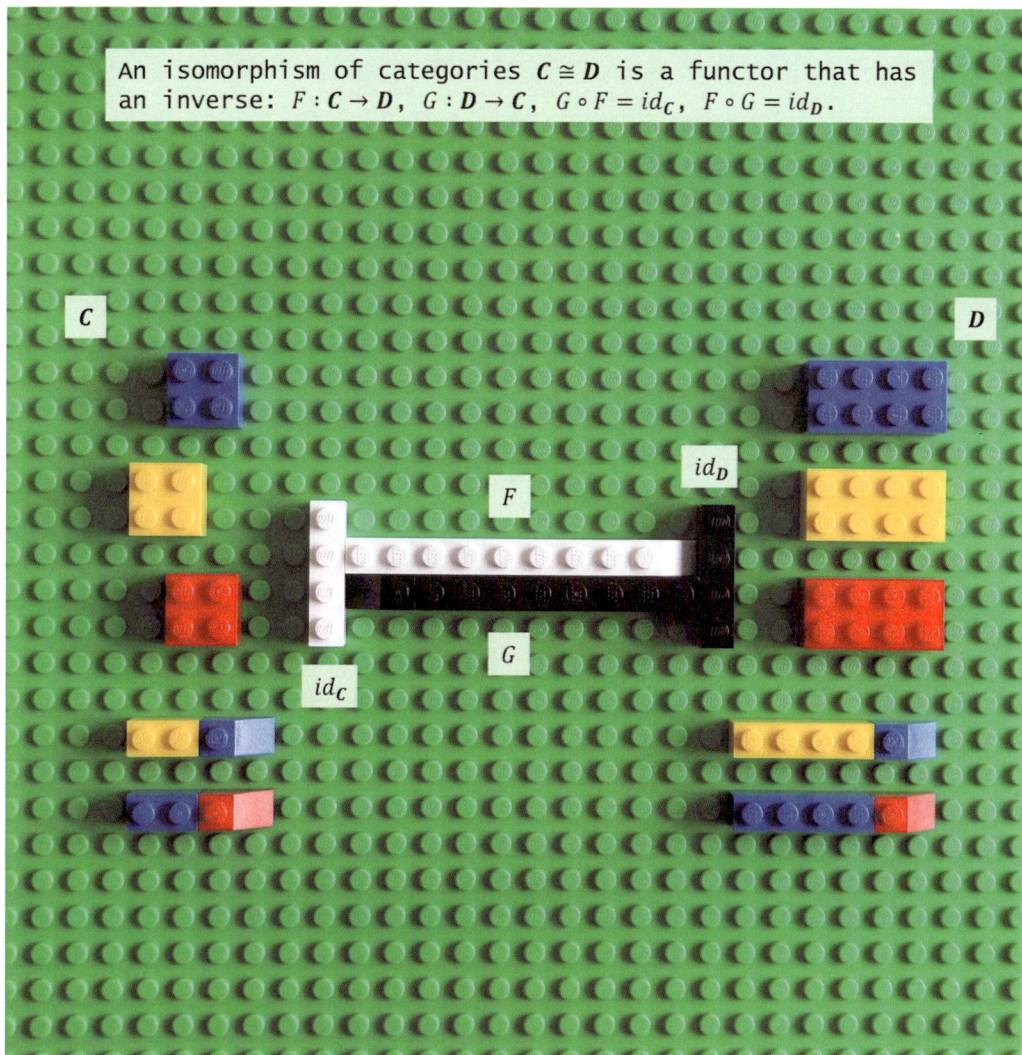

An isomorphism of categories $C \cong D$ is a functor that has an inverse: $F : C \to D$, $G : D \to C$, $G \circ F = id_C$, $F \circ G = id_D$.

C

D

id_D

F

G

id_C

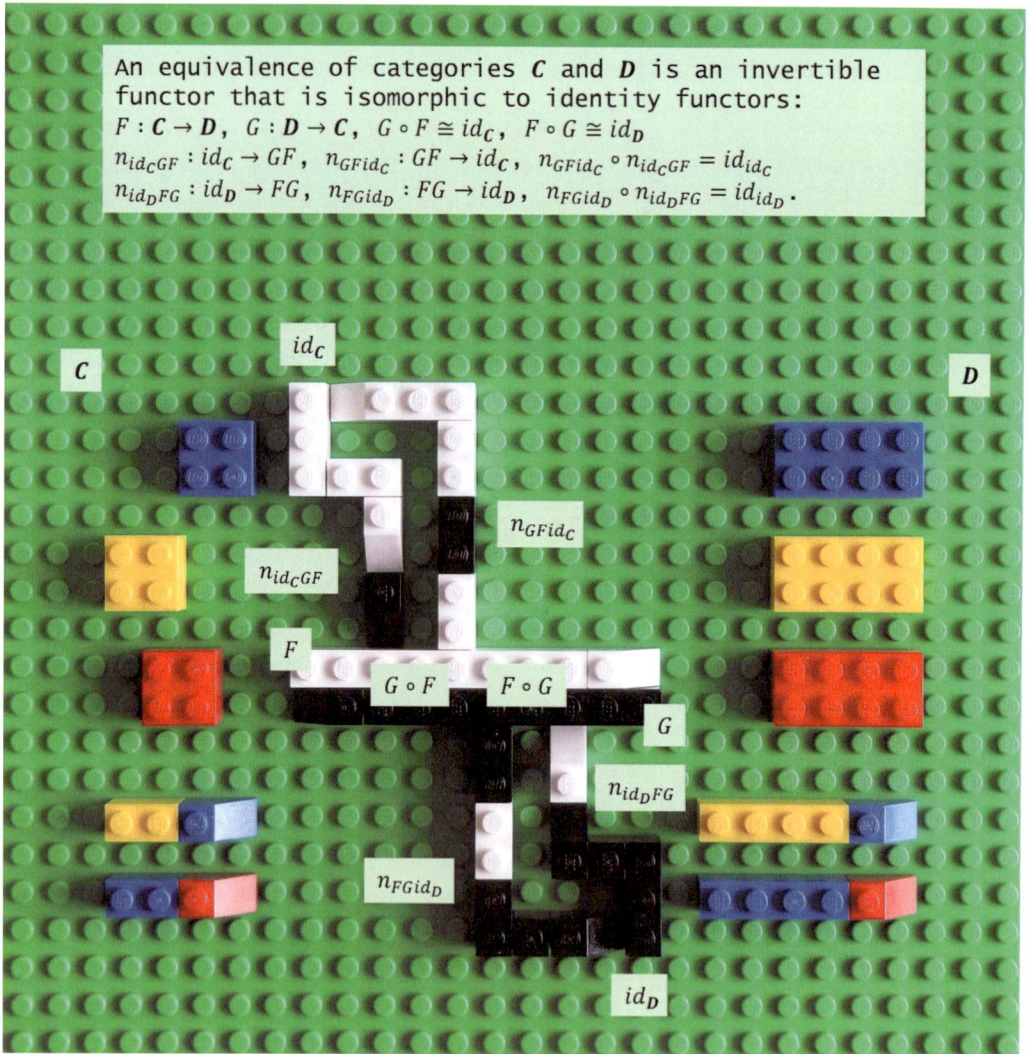

An equivalence of categories **C** and **D** is an invertible functor that is isomorphic to identity functors:
$F : \mathbf{C} \to \mathbf{D}$, $G : \mathbf{D} \to \mathbf{C}$, $G \circ F \cong id_{\mathbf{C}}$, $F \circ G \cong id_{\mathbf{D}}$
$n_{id_{\mathbf{C}}GF} : id_{\mathbf{C}} \to GF$, $n_{GFid_{\mathbf{C}}} : GF \to id_{\mathbf{C}}$, $n_{GFid_{\mathbf{C}}} \circ n_{id_{\mathbf{C}}GF} = id_{id_{\mathbf{C}}}$
$n_{id_{\mathbf{D}}FG} : id_{\mathbf{D}} \to FG$, $n_{FGid_{\mathbf{D}}} : FG \to id_{\mathbf{D}}$, $n_{FGid_{\mathbf{D}}} \circ n_{id_{\mathbf{D}}FG} = id_{id_{\mathbf{D}}}$.

C

D

$id_{\mathbf{C}}$

$n_{GFid_{\mathbf{C}}}$

$n_{id_{\mathbf{C}}GF}$

F

$G \circ F$ $F \circ G$

G

$n_{id_{\mathbf{D}}FG}$

$n_{FGid_{\mathbf{D}}}$

$id_{\mathbf{D}}$

Part 6 page 8

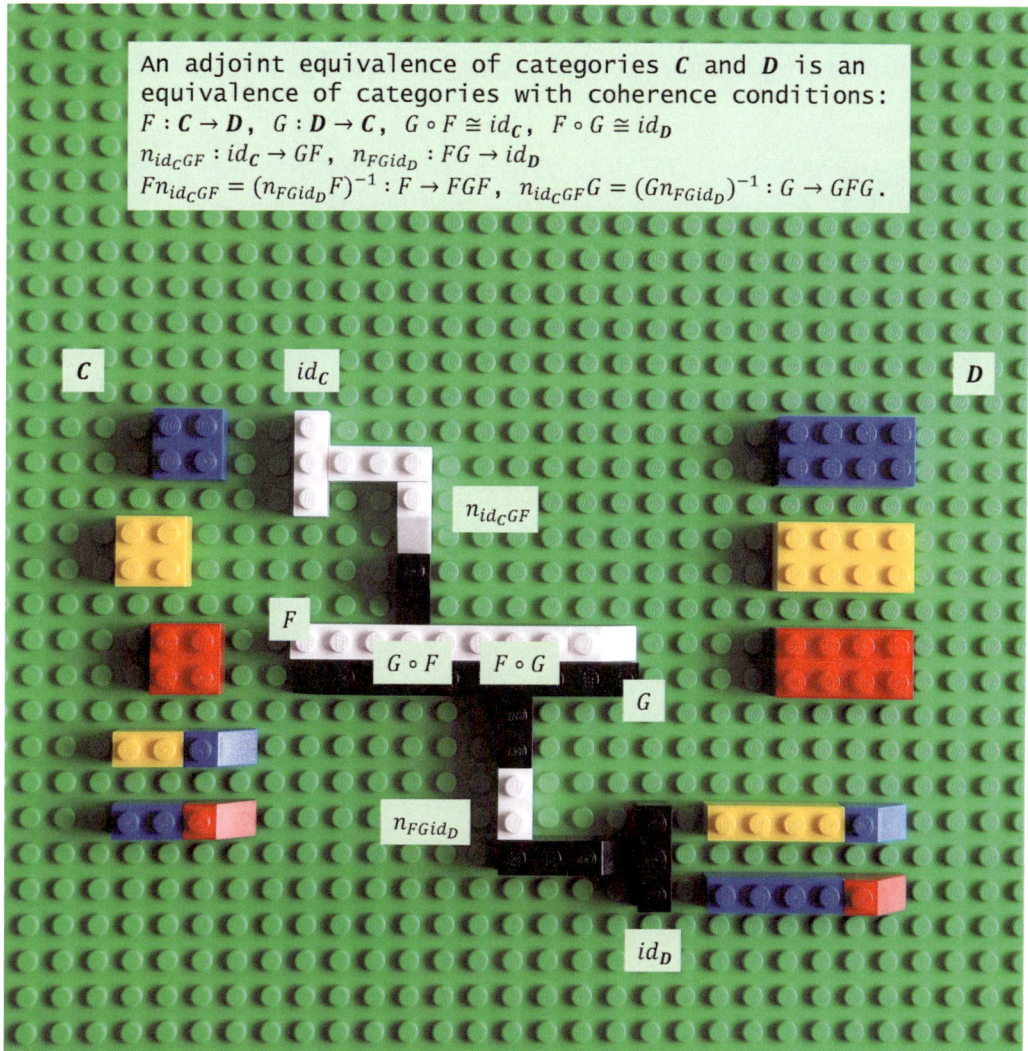

An adjoint equivalence of categories C and D is an equivalence of categories with coherence conditions:
$F : C \to D$, $G : D \to C$, $G \circ F \cong id_C$, $F \circ G \cong id_D$
$n_{id_C GF} : id_C \to GF$, $n_{FGid_D} : FG \to id_D$
$Fn_{id_C GF} = (n_{FGid_D}F)^{-1} : F \to FGF$, $n_{id_C GF}G = (Gn_{FGid_D})^{-1} : G \to GFG$.

C id_C D

$n_{id_C GF}$

F

$G \circ F$ $F \circ G$

G

n_{FGid_D}

id_D

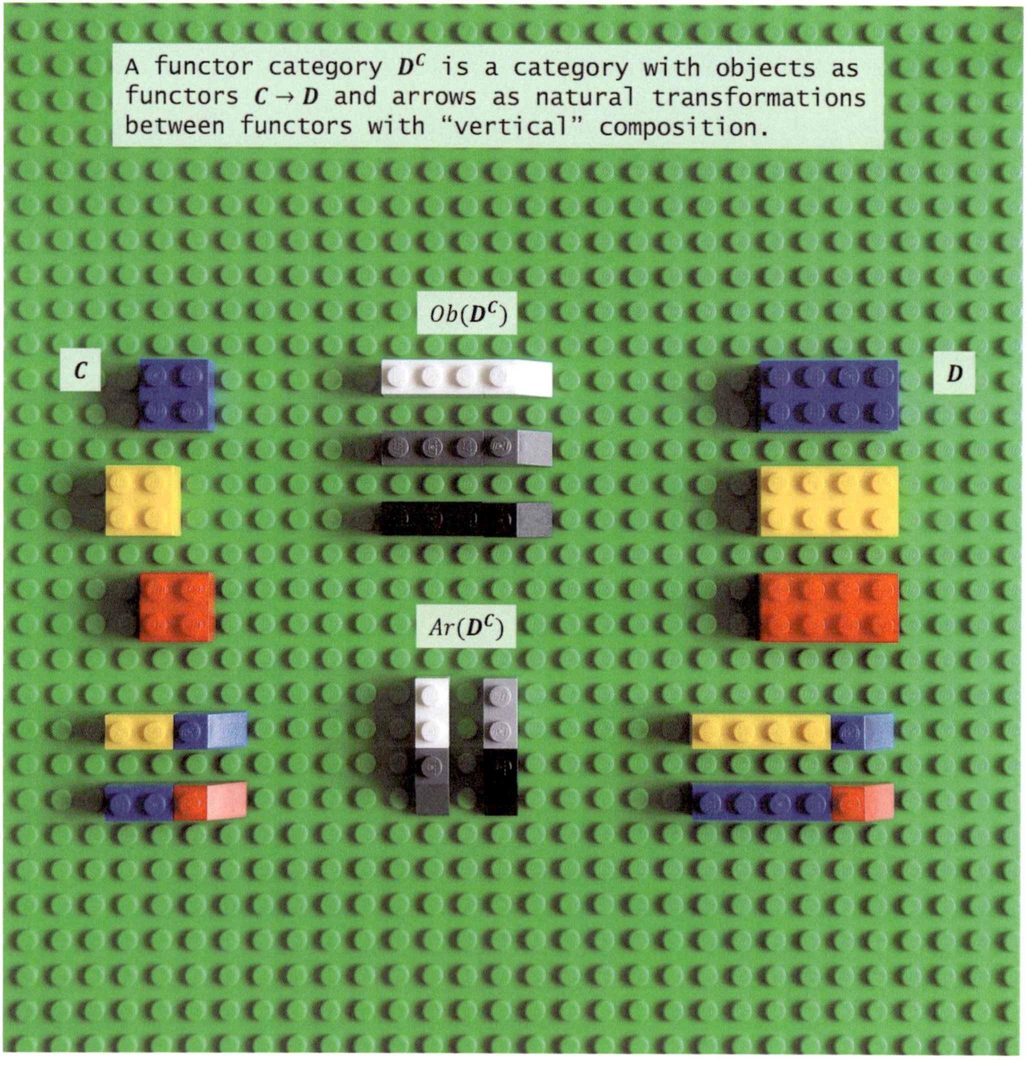

A functor category D^C is a category with objects as functors $C \to D$ and arrows as natural transformations between functors with "vertical" composition.

$Ob(D^C)$

C

D

$Ar(D^C)$

2 is a category with $Ob(\mathbf{2}) = \{0,1\}$, $Ar(\mathbf{2}) = \{0 \to 1, id_0, id_1\}$.
$Ar(\mathbf{C^2}) \ni Ar(F,G)$ is a natural transformation between $Ob(\mathbf{C^2}) \ni F,G : \mathbf{2} \to \mathbf{C}$.
$Ar(F,G) = Ar(F(0),G(0)) \times Ar(F(1),G(1))$ is an arrow between arrows in $\mathbf{C}$.
$\mathbf{C^2}$ is a category of morphisms (arrows) of $\mathbf{C}$.
Also compare with a natural transformation picture in $Part_1$.

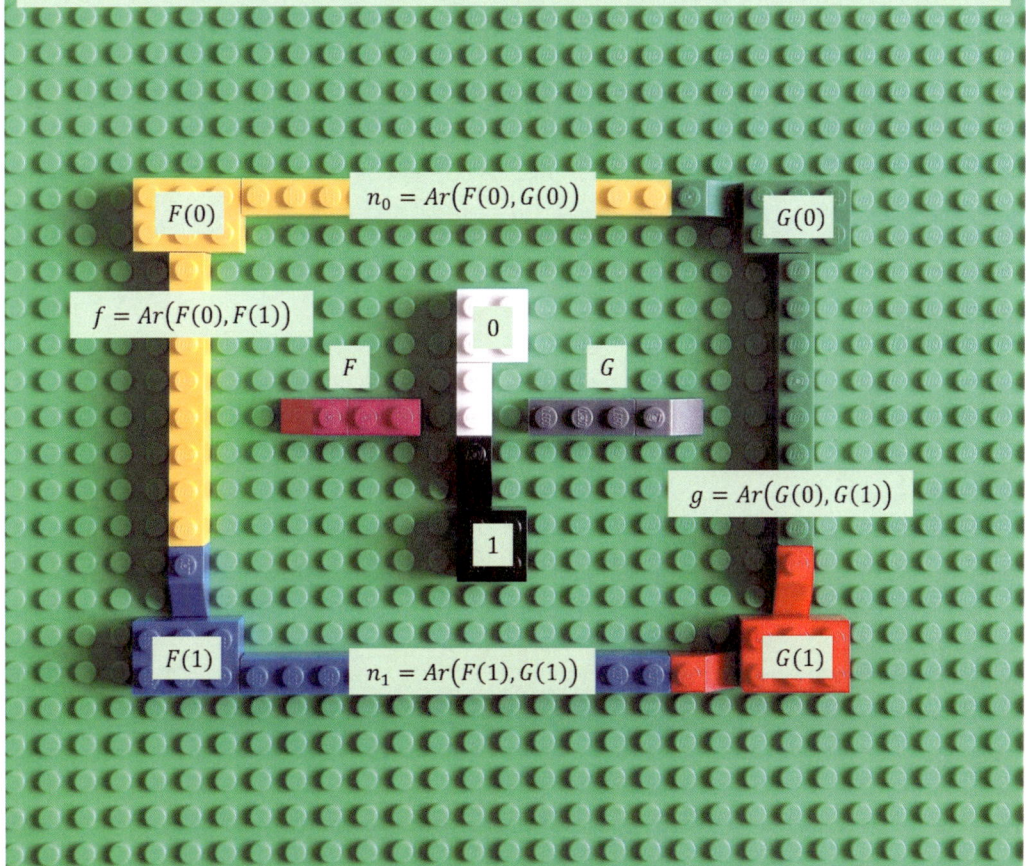

A natural transformation can be represented as functors
$C \times 2 \to D$ and $C \to D^2$.

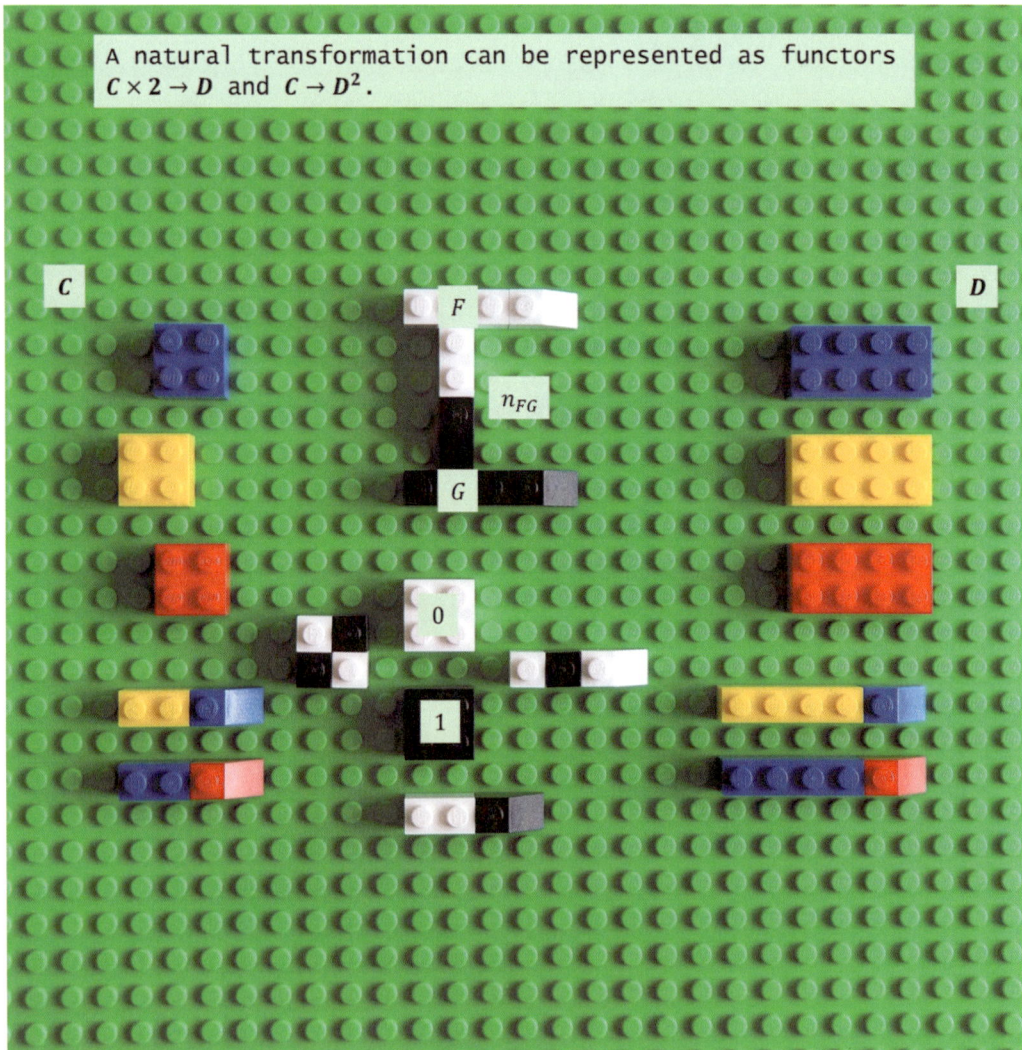

A covariant representable functor $R: \boldsymbol{C} \to \boldsymbol{Set}$ is isomorphic to a covariant morphism functor (introduced in $Part_5$): $R \cong F(X_0, -)$.
$\forall X \in Ob(\boldsymbol{C}) \; \exists X_0 \in Ob(\boldsymbol{C}) \; R(X) \cong F(X_0, X) = R_{X_0}(X)$

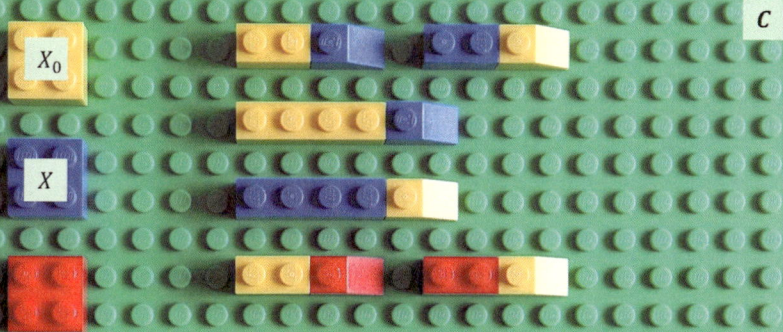

$\boldsymbol{C}$

X_0

X

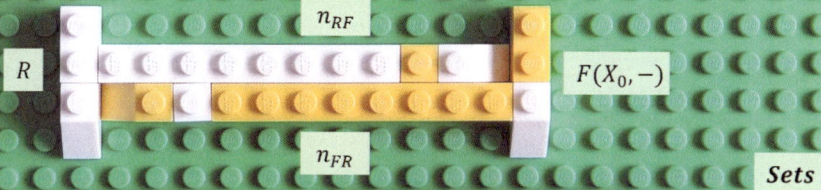

n_{RF}

R — $F(X_0, -)$

n_{FR}

$\boldsymbol{Sets}$

$R(X)$

$F(X_0, X)$

A contravariant morphism functor $F(-, X_0) : C^{op} \to \textbf{Sets}$
sends $X \in Ob(C)$ to $\{f | f \in Ar(X, X_0)\}$.

X_0

X

Y

$F(-, X_0)$

$\{f | f \in Ar(X, X_0)\}$ $\{g | g \in Ar(Y, X_0)\}$

 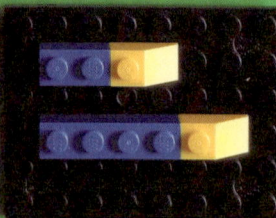

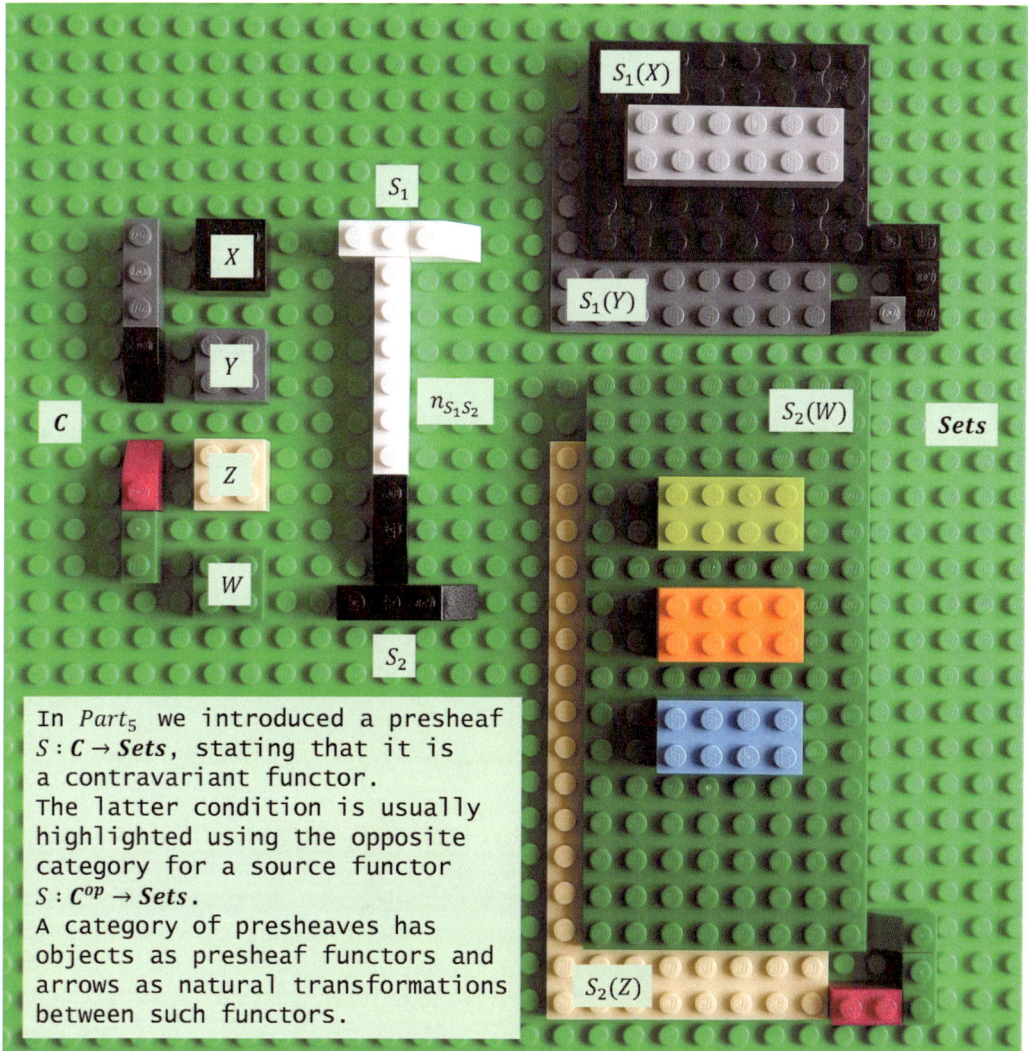

$S_1(X)$

S_1

X

$S_1(Y)$

Y

C

$n_{S_1 S_2}$

$S_2(W)$

Sets

Z

W

S_2

$S_2(Z)$

In $Part_5$ we introduced a presheaf
$S : C \rightarrow Sets$, stating that it is
a contravariant functor.
The latter condition is usually
highlighted using the opposite
category for a source functor
$S : C^{op} \rightarrow Sets$.
A category of presheaves has
objects as presheaf functors and
arrows as natural transformations
between such functors.

A contravariant representable functor $R : C^{op} \to Set$ (representable presheaf) is isomorphic to a contravariant morphism functor: $R \cong F(-, X_0)$. Representable functors form a representable functor category $\boldsymbol{R} = \boldsymbol{Set}^{C^{op}}$ (a category of representable presheaves).
$\forall X \in Ob(\boldsymbol{C}) \; \exists X_0 \in Ob(\boldsymbol{C}) \; R(X) \cong F(X, X_0) = R_{X_0}(X)$

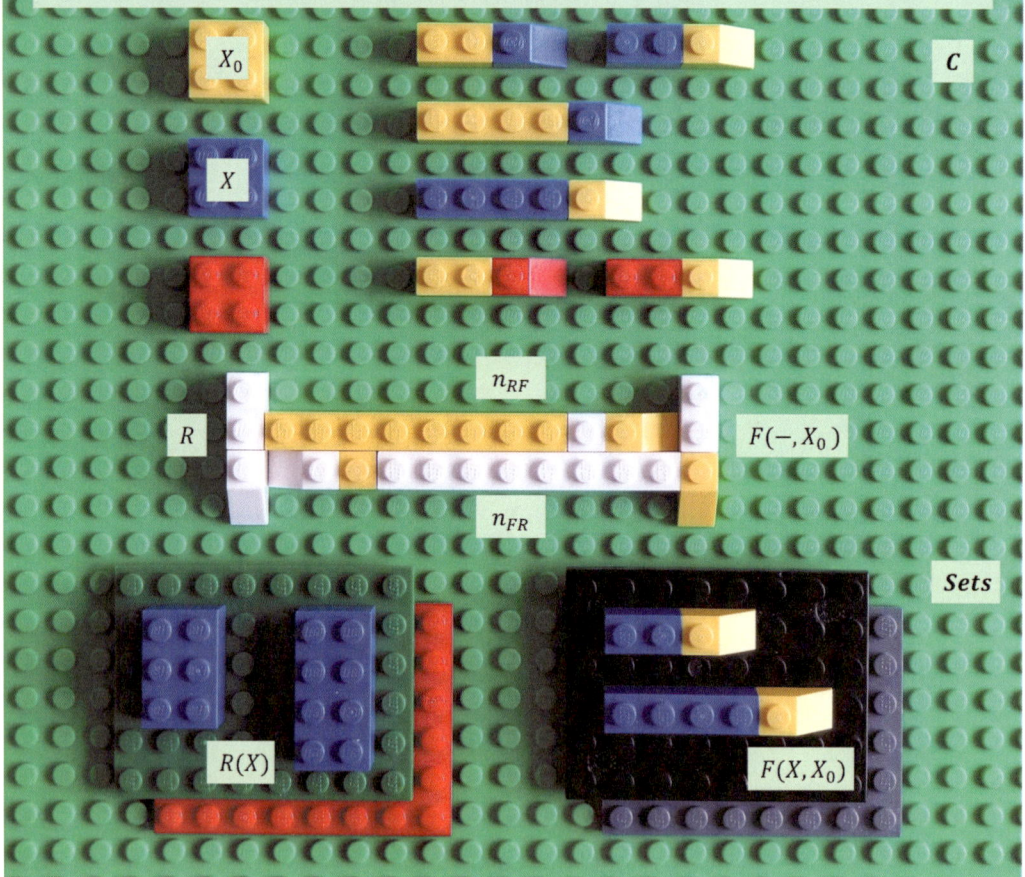

X_0

C

X

n_{RF}

R $F(-, X_0)$

n_{FR}

Sets

$R(X)$

$F(X, X_0)$

The Yoneda embedding is a functor $Y_{em} : C \to R$ from a category C to a category of its contravariant representable functors (representable presheaves): $Y_{em} : C \to Set^{C^{op}}$.

X

Y

$R_X = Y_{em}(X)$ $R_Y = Y_{em}(Y)$

We rearrange objects of C from the previous page to show C object mapping. The Yoneda embedding $C \to Set^{C^{op}}$ maps $X \in Ob(C)$ to a representable presheaf R_X, and maps $f \in Ar(C)$ to a corresponding natural transformation between target functors.

X f Y

Y_{em}

$R_X = Y_{em}(X)$

$n_{R_X R_Y} = Y_{em}(f)$

$R_Y = Y_{em}(Y)$

The Yoneda lemma states that $\forall X \in Ob(C)$ and $\forall S \in C^{op} \to Sets$ (presheaf) there is an isomorphism between $S(X)$ and $Ar(Y_{em}(X), S)$, the set of arrows between the Yoneda embedding of X (representable presheaf) and S.

X

Y_{em}

S $n_{R_X S}$ $R_X = Y_{em}(X)$

$\cong$

$S(X)$

Index for Part$_1$-Part$_6$

Visual Category Theory

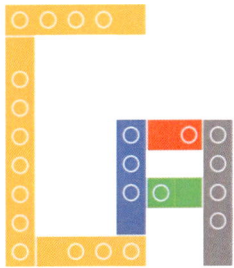

Dmitry Vostokov

brick by brick

Part$_7 \in$ C^{Parts}

Visual Category Theory Brick by Brick, Part 7: Using LEGO® to Teach Abstract Mathematics

Published by OpenTask, Republic of Ireland

OpenTask books and magazines are available through booksellers and distributors worldwide. For further information or comments, send requests to press@opentask.com.

A CIP catalog record for this book is available from the British Library.

ISBN-13: 978-1912636464 (Paperback)

Revision 1.00 (December 2020)

In $Part_5$ we introduced an exponential of objects $X, Y \in Ob(\pmb{C})$. It can be thought of as representing a collection of arrows between two objects X and Y. It is also an object of $\pmb{C}$ (internal view) in contrast to a set of arrows $Hom_C(X,Y) = \{f \mid f \in Ar(X,Y)\} \in \pmb{Sets}$, also called hom-set (external view).

$X, Y, Y^X \in Ob(\pmb{C})$

Y^X

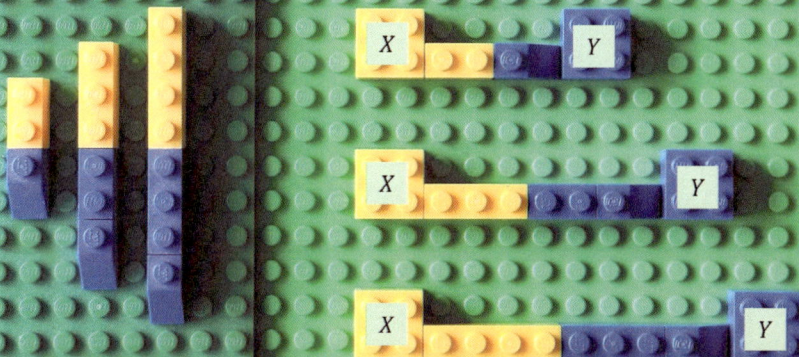

In addition to 14 books mentioned in $Part_1$, $Part_3$ - $Part_6$ we also used:

- Enriched Meanings: Natural Language Semantics with Category Theory by Ash Asudeh and Gianluca Giorgolo
- Categories for the Working Mathematician, Second Edition by Saunders Mac Lane

In $Part_1$ we introduced functors between categories C and D.
If C and D are the same category, we have endofunctors $F : C \to C$,
$X, Y, W, Z \in Ob(C)$, $f, g \in Ar(C)$.

The disjoint union of an indexed family of sets $\{S_i | i \in I\}$ is the set of pairs of individual members with indexes $\bigcup_{i \in I} \{(x, i) | x \in S_i\}$.

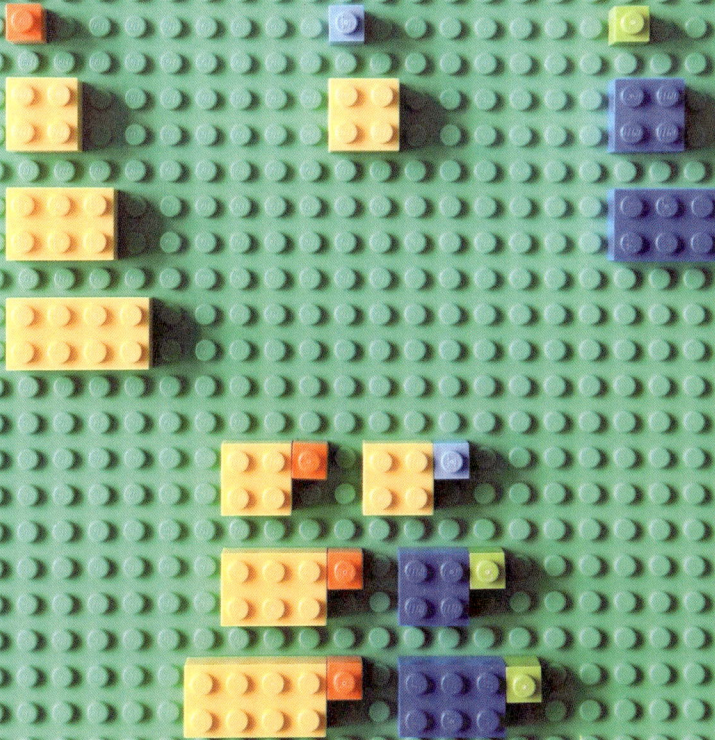

A partial function $f: X \rightarrow Y$ does not map some of its domain elements to its codomain. We say that the function is not defined for such elements.

We can make the function f total f^T by extending its codomain Y with a distinguished element d to which we send elements for which the function f is undefined. The codomain of f^T is a disjoint union of the codomain Y of f and $\{d\}$: $f^T\colon X \to Y \sqcup \{d\}$.

X f^T $Y \sqcup \{d\}$

d

A composition of a partial function $f: X \to Y$ and $g: Y \to Z$ is only defined for values of X which are mapped to values of Y for which the function g is defined. How do we make the function composition a total function?

X   f   Y   g   Z

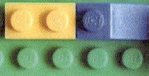

We first make the function f total f^T by extending its codomain Y with a distinguished element d.

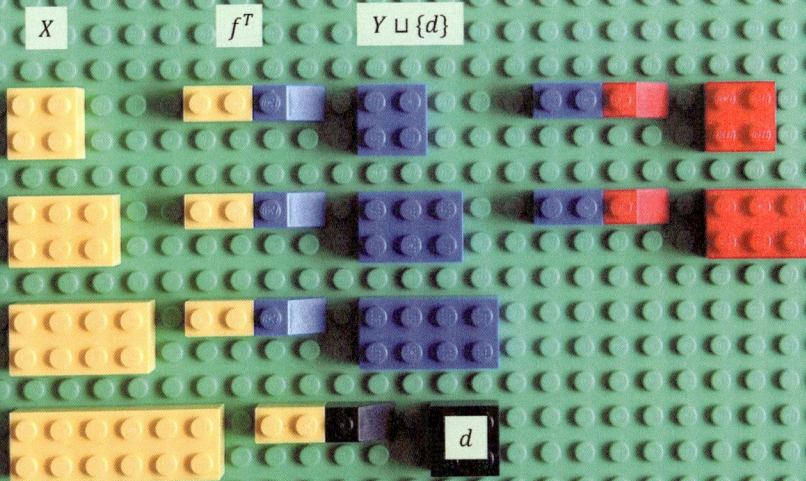

X $\qquad$ f^T $\qquad$ $Y \sqcup \{d\}$

d

$n_Y : Y \to Y \sqcup \{d\}$ is an embedding transformation of the codomain of f into its disjoint union with a distinguished element d.

n_Y

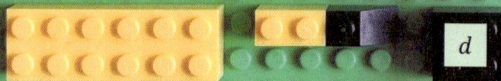

d

$Y \sqcup \{d\}$

$n_Y: Y \to Y \sqcup \{d\}$ transformation of the codomain can be abstracted in category theory as a natural transformation (introduced in $Part_1$) between identity and $T: C \to C$ endofunctors in the category of sets as objects and functions as arrows. T maps objects to their disjoint union with a distinguished element d, and arrows $X \to Y$ are mapped to $X \sqcup \{d\} \to Y \sqcup \{d\}$.

We also make the function g total g^T by extending its codomain Z with a distinguished element d.

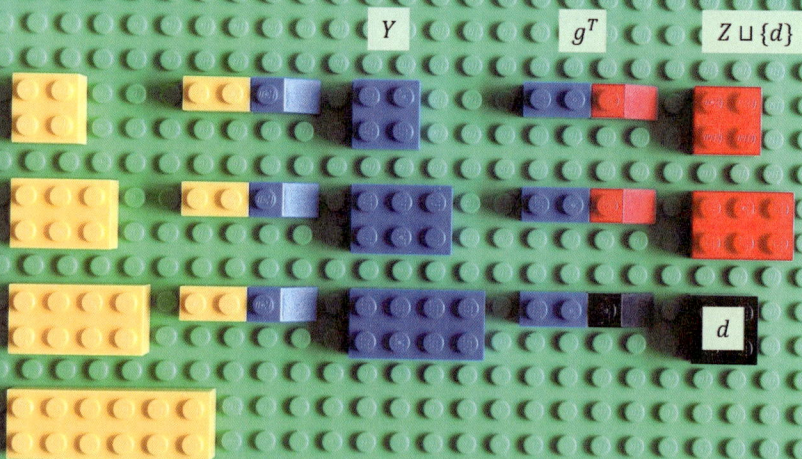

$n_Z: Z \to Z \sqcup \{d\}$ is an embedding transformation of the codomain of g into its disjoint union with a distinguished element d.

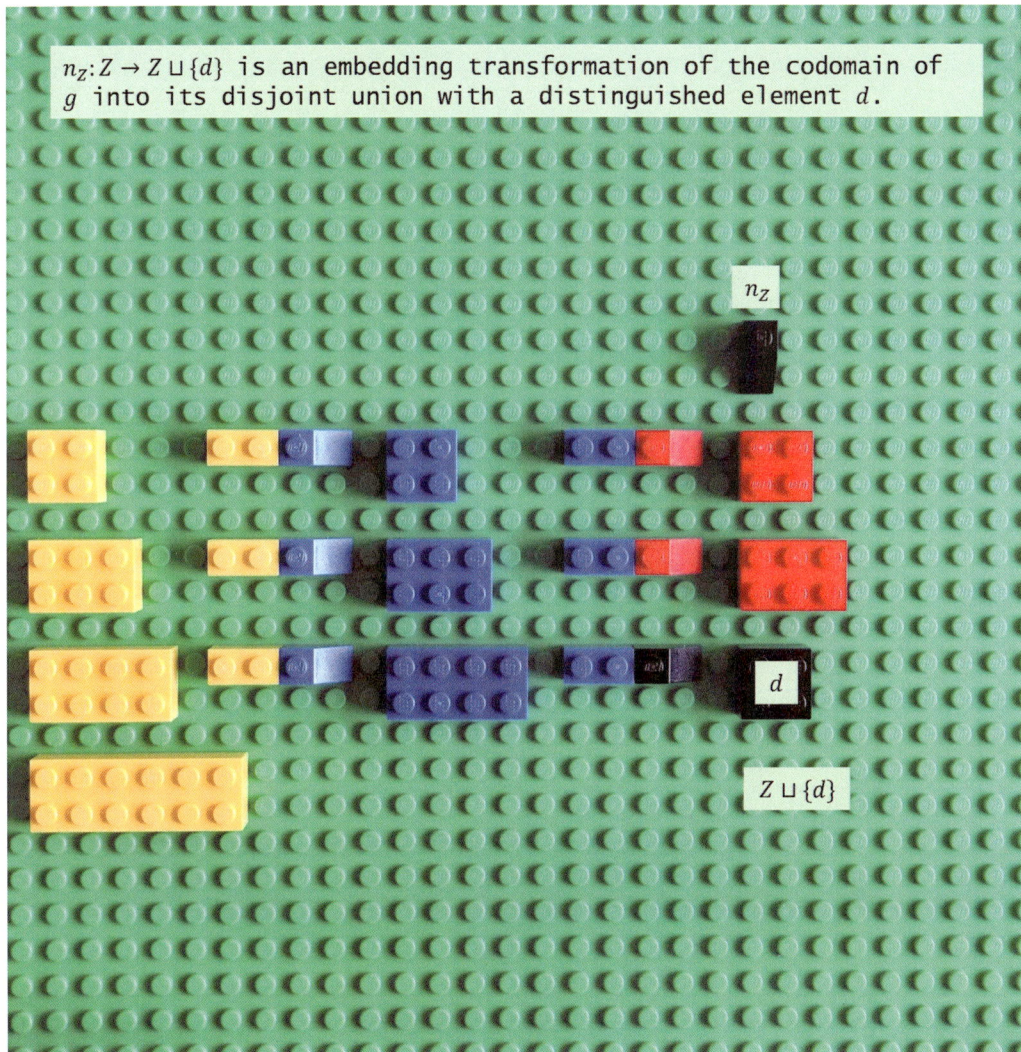

n_Z

d

$Z \sqcup \{d\}$

$n_Z : Z \to Z \sqcup \{d\}$ transformation of the codomain can be abstracted in category theory as a natural transformation between identity and $T : C \to C$ endofunctors. T maps objects to their disjoint union with a distinguished element d, and arrows $Y \to Z$ are mapped to $Y \sqcup \{d\} \to Z \sqcup \{d\}$.

We now want to compose the new function f^T with
the new function g^T to have a total function.

X f^T $Y \sqcup \{d\}$ g^T $Z \sqcup \{d\}$

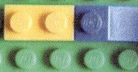

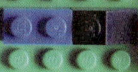

We also have $n_Y: Y \to Y \sqcup \{d\}$ and $n_Z: Z \to Z \sqcup \{d\}$ embedding transformations of the respective codomains into their disjoint unions with a distinguished element d.

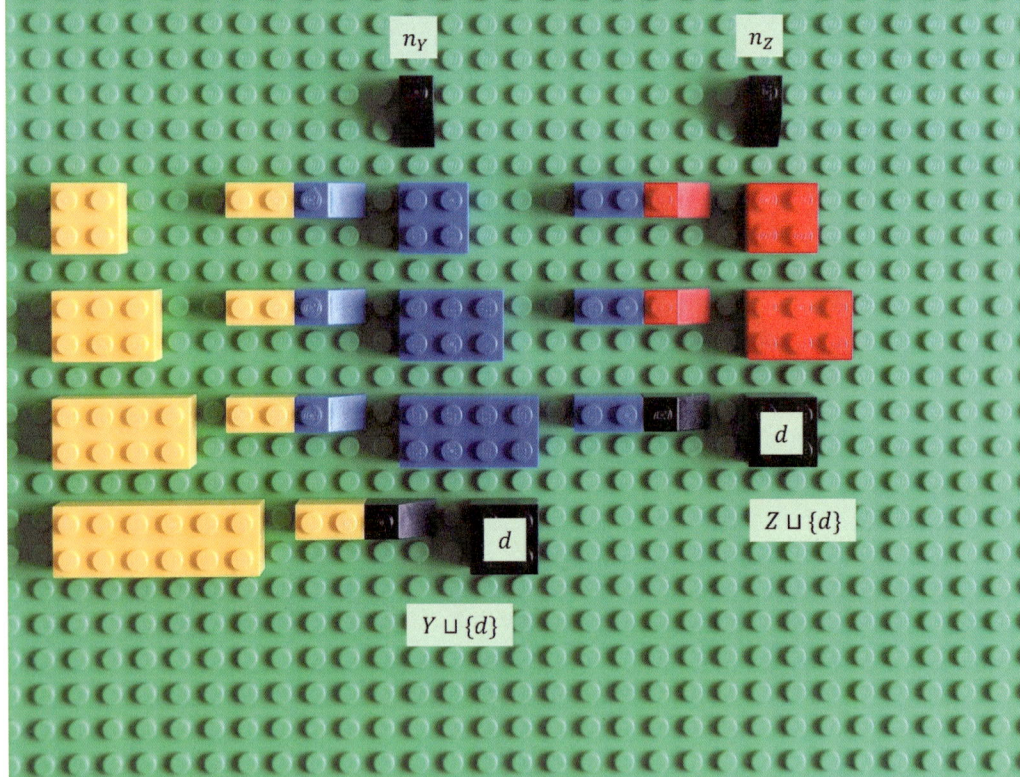

n_Y

n_Z

d

$Z \sqcup \{d\}$

d

$Y \sqcup \{d\}$

We extend the codomain $Z \sqcup \{d\}$ with the same distinguished element d by making a disjoint union again. We now have the new total function $g^{TT} : Y \sqcup \{d\} \to Z \sqcup \{d\} \sqcup \{d\}$.

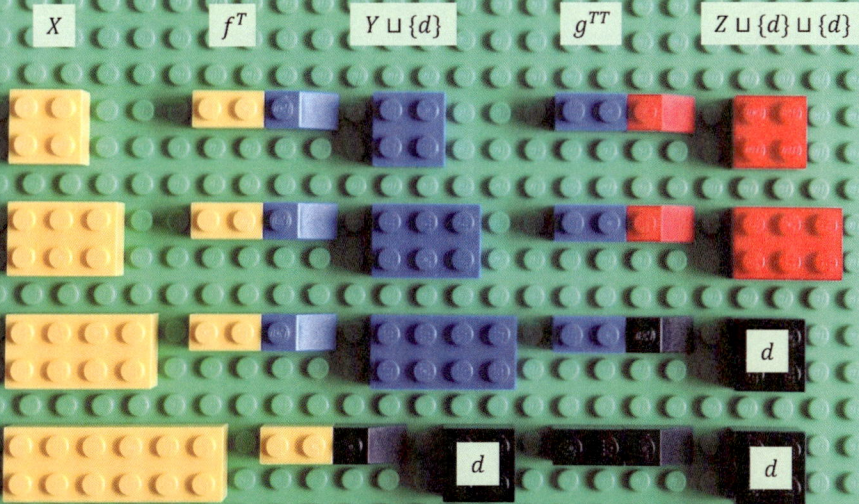

X f^T $Y \sqcup \{d\}$ g^{TT} $Z \sqcup \{d\} \sqcup \{d\}$

$n_{Z \sqcup \{d\}}: Z \sqcup \{d\} \to Z \sqcup \{d\} \sqcup \{d\}$ is an embedding transformation of the codomain of g^{TT} into its disjoint union with a distinguished element d. We now have a function $g^{TT} \circ f^T: X \to Z \sqcup \{d\} \sqcup \{d\}$, but we need a total function $X \to Z \sqcup \{d\}$.

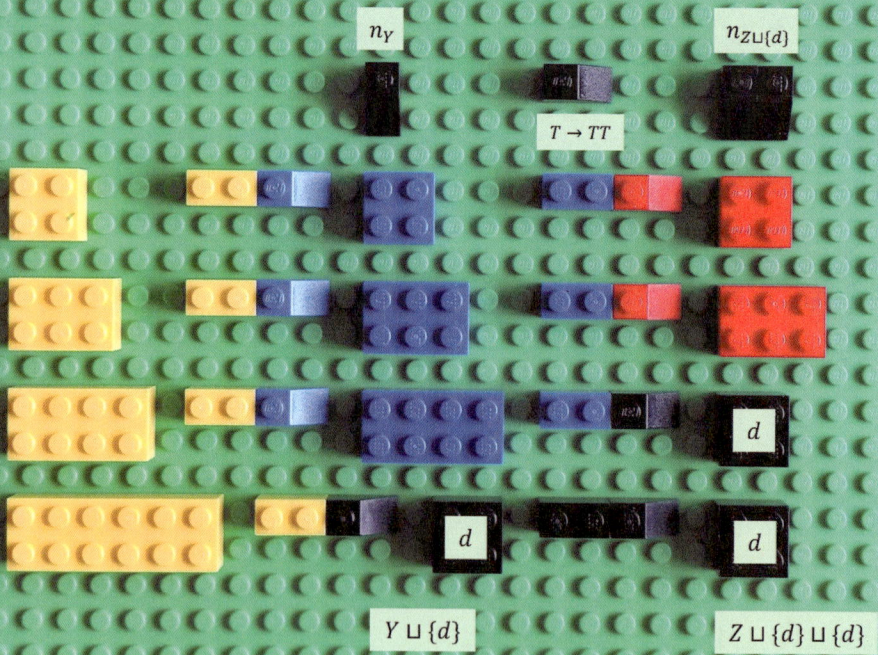

$n_{Z \sqcup \{d\}} : Z \sqcup \{d\} \to Z \sqcup \{d\} \sqcup \{d\}$ transformation of the codomain can be abstracted in category theory as a natural transformation between identity and $T : \boldsymbol{C} \to \boldsymbol{C}$ endofunctors. T maps objects to their disjoint union with a distinguished element d, and arrows $Y \sqcup \{d\} \to Z \sqcup \{d\}$ are mapped to $Y \sqcup \{d\} \sqcup \{d\} \to Z \sqcup \{d\} \sqcup \{d\}$.

$\{d\} \sqcup Y$

$\{d\} \sqcup Y$

id_C

$\{d\} \sqcup Z$

T

$Y \sqcup \{d\} \sqcup \{d\}$

$\{d\} \sqcup Z$

$Z \sqcup \{d\} \sqcup \{d\}$

$n_{Z \sqcup \{d\}}$

We introduce a collapsing function $m: Z \sqcup \{d\} \sqcup \{d\} \to Z \sqcup \{d\}$.

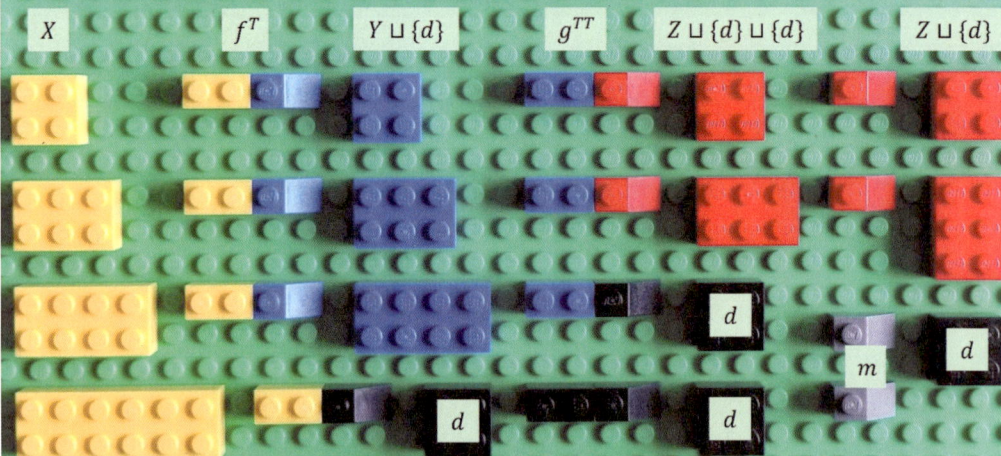

X f^T $Y \sqcup \{d\}$ g^{TT} $Z \sqcup \{d\} \sqcup \{d\}$ $Z \sqcup \{d\}$

$m_{Z \sqcup \{d\} \sqcup \{d\}} : Z \sqcup \{d\} \sqcup \{d\} \to Z \sqcup \{d\}$ is a transformation of the codomain of the double transformation $n \circ n$ into the codomain of one such transformation n.

We now have a total function $m_{Z \sqcup \{d\} \sqcup \{d\}} \circ g^{TT} \circ f^T : X \to Z \sqcup \{d\}$.

n_Y $n_{Z \sqcup \{d\}}$ $m_{Z \sqcup \{d\} \sqcup \{d\}}$

$T \to TT$ $TT \to T$

d

d

d d $Z \sqcup \{d\}$

$Y \sqcup \{d\}$ $Z \sqcup \{d\} \sqcup \{d\}$

m transformation of the double transformation $n \circ n$ into the codomain of one such transformation n can be abstracted in category theory as natural transformations between $TT: \textbf{\textit{C}} \to \textbf{\textit{C}}$ and $T: \textbf{\textit{C}} \to \textbf{\textit{C}}$ endofunctors.

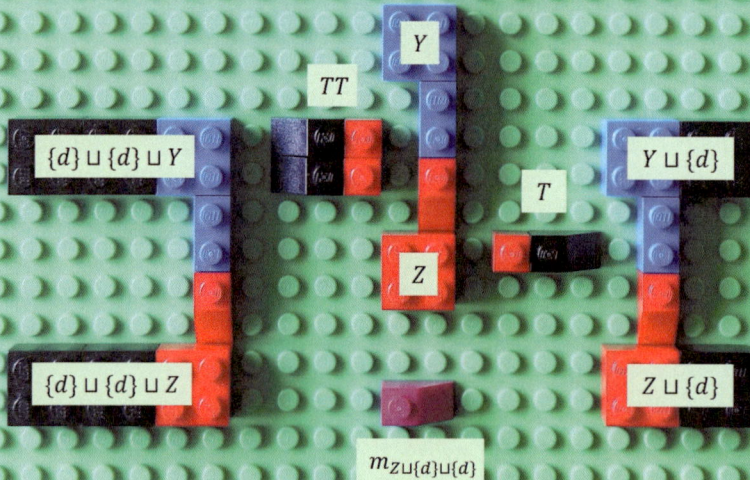

A monad in category theory is a triple $\{T, n, m\}$ where T is an endofunctor, n, m are natural transformations with the following condition: $m_{TTZ} \circ n_{TZ} \circ T = m_{TTZ} \circ T \circ n_Z = id_T$ (identity transformation between T endofunctors).
n can be considered as a monad identity element.
Note: in $Part_6$ we introduced composition of natural transformations and functors.

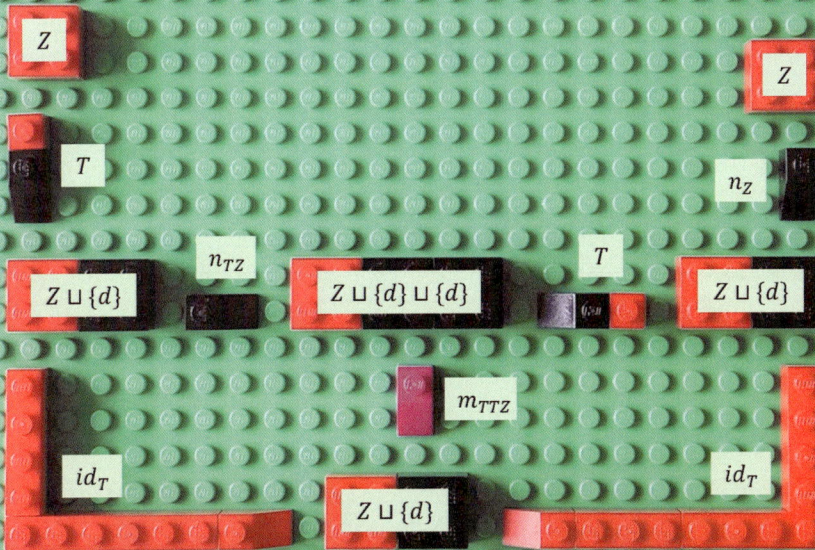

Z

Z

T

n_Z

n_{TZ}

T

$Z \sqcup \{d\}$

$Z \sqcup \{d\} \sqcup \{d\}$

$Z \sqcup \{d\}$

m_{TTZ}

id_T

id_T

$Z \sqcup \{d\}$

$n_{TZ} \circ T = T \circ n_z$ brick illustration can be rebuilt using disjoint union notation.

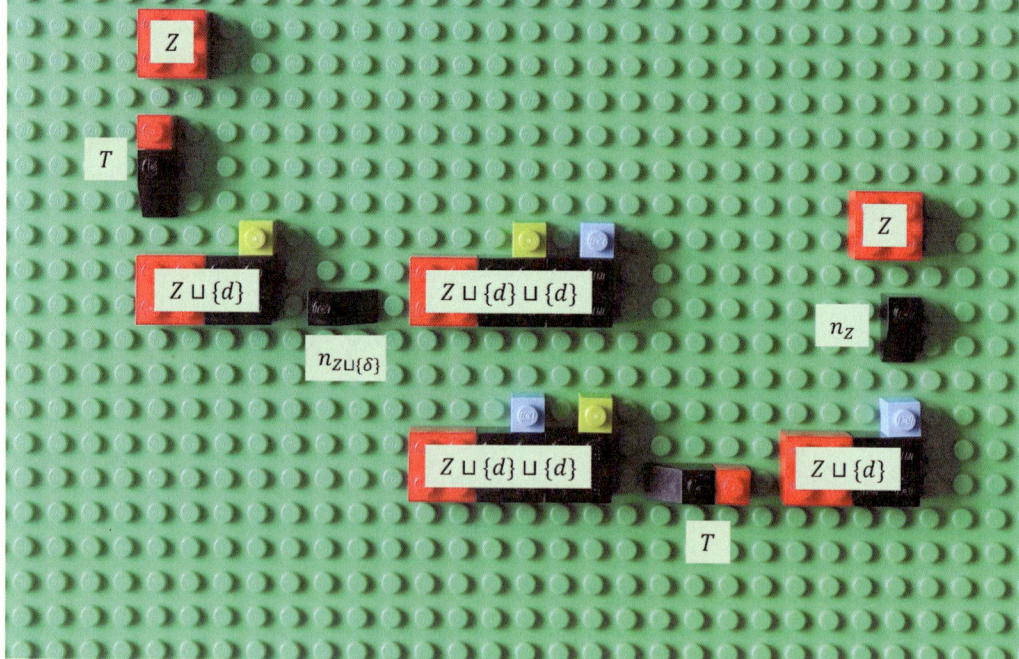

Z

T

$Z \sqcup \{d\}$

$n_{Z \sqcup \{\delta\}}$

$Z \sqcup \{d\} \sqcup \{d\}$

Z

n_z

$Z \sqcup \{d\} \sqcup \{d\}$

T

$Z \sqcup \{d\}$

Another monad condition is related to m natural transformation: $m_{TTZ} \circ m_{TTTZ} = m_{TTZ} \circ m_{TT(TZ)} = m_{TTZ} \circ m_{T(TT)Z}$ where we can "multiply" outer or inner layers. Therefore, m can be considered a binary operation. A monad on C can also be defined as a monoid (introduced in $Part_4$) in the category of endofunctors of C, End_C, where morphisms are natural transformations between endofunctors.

$m_{TT(TZ)}$

$m_{T(TT)Z}$

m_{TTZ}

m_{TTZ}

www.ingramcontent.com/pod-product-compliance
Lightning Source LLC
Chambersburg PA
CBRC091940210326
41598CB00013B/874